A-level Study Guide

Chemistry

Revised and updated for 2008 by

Philip Barratt

Michael Cox

Revision Express

Series Consultants: Geoff Black and Stuart Wall

Project Manager: Hywel Evans

Pearson Education Limited

Edinburgh Gate, Harlow

Essex CM20 2JE, England

and Associated Companies throughout the world

© Pearson Education Limited 2000, 2003, 2008

British Library Cataloguing in Publication Data

A catalogue entry for this title is available from the British Library.

ISBN 978-1-4082-0651-5

First published 2000
Reprinted 2001
Updated 2003, 2008
New Edition 2008
Third impression 2010

Set by Juice Creative Ltd

Printed and bound in China SWTC/03

Contents

How to use this book

Specification map
Provides a quick and easy overview of the topic that you need to study for the specification you are studying.

Exam skills
Gives a quick overview of the key skills that will be tested in the exam

By the end of this chapter you will be able to

- Write and balance different types of chemical equation and work out empirical formulae using masses or percentage composition by mass

Don't forget
Flags up areas of study that you must remember and which students commonly forget

Action point
A suggested activity linked to the content

Checkpoint
Quick question to check your understanding with full answers given at the end of the chapter

Links
Cross-reference links to other relevant sections in the book

Examiner's secrets
Hints and tips for exam success

Watch out!
Flags up common mistakes and gives hints on how to avoid them

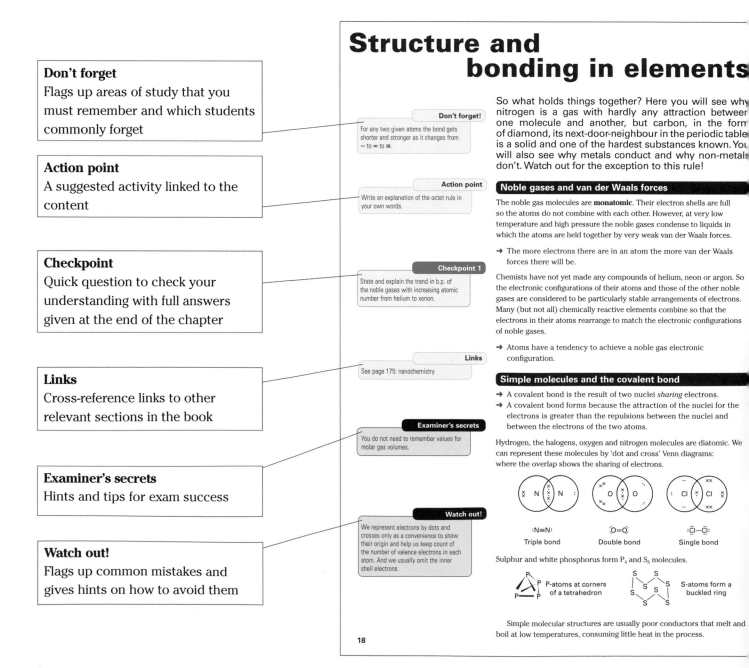

Structure and bonding in elements

So what holds things together? Here you will see why nitrogen is a gas with hardly any attraction between one molecule and another, but carbon, in the form of diamond, its next-door-neighbour in the periodic table, is a solid and one of the hardest substances known. You will also see why metals conduct and why non-metals don't. Watch out for the exception to this rule!

Don't forget!
For any two given atoms the bond gets shorter and stronger as it changes from — to ═ to ≡.

Action point
Write an explanation of the octet rule in your own words.

Checkpoint 1
State and explain the trend in b.p. of the noble gases with increasing atomic number from helium to xenon.

Links
See page 175: nanochemistry

Examiner's secrets
You do not need to remember values for molar gas volumes.

Watch out!
We represent electrons by dots and crosses only as a convenience to show their origin and help us keep count of the number of valence electrons in each atom. And we usually omit the inner shell electrons.

Noble gases and van der Waals forces

The noble gas molecules are **monatomic**. Their electron shells are full so the atoms do not combine with each other. However, at very low temperature and high pressure the noble gases condense to liquids in which the atoms are held together by very weak van der Waals forces.

→ The more electrons there are in an atom the more van der Waals forces there will be.

Chemists have not yet made any compounds of helium, neon or argon. So the electronic configurations of their atoms and those of the other noble gases are considered to be particularly stable arrangements of electrons. Many (but not all) chemically reactive elements combine so that the electrons in their atoms rearrange to match the electronic configurations of noble gases.

→ Atoms have a tendency to achieve a noble gas electronic configuration.

Simple molecules and the covalent bond

→ A covalent bond is the result of two nuclei *sharing* electrons.
→ A covalent bond forms because the attraction of the nuclei for the electrons is greater than the repulsions between the nuclei and between the electrons of the two atoms.

Hydrogen, the halogens, oxygen and nitrogen molecules are diatomic. We can represent these molecules by 'dot and cross' Venn diagrams: where the overlap shows the sharing of electrons.

:N≡N:　　　　:O=O:　　　　:C̈l–C̈l:
Triple bond　　Double bond　　Single bond

Sulphur and white phosphorus form P_4 and S_8 molecules.

P-atoms at corners of a tetrahedron

S-atoms form a buckled ring

Simple molecular structures are usually poor conductors that melt and boil at low temperatures, consuming little heat in the process.

Topic checklist

A topic overview of the content covered and how it matches to the specification you are studying

Topic checklist

	Edexcel		AQA		OCR		WJEC		CCEA	
	AS	A2	AS	A2	AS	A2	AS	A2	AS	A2
Formulae and equations	O		O		O		O		O	
Atoms and isotopes	O		O		O		O		O	

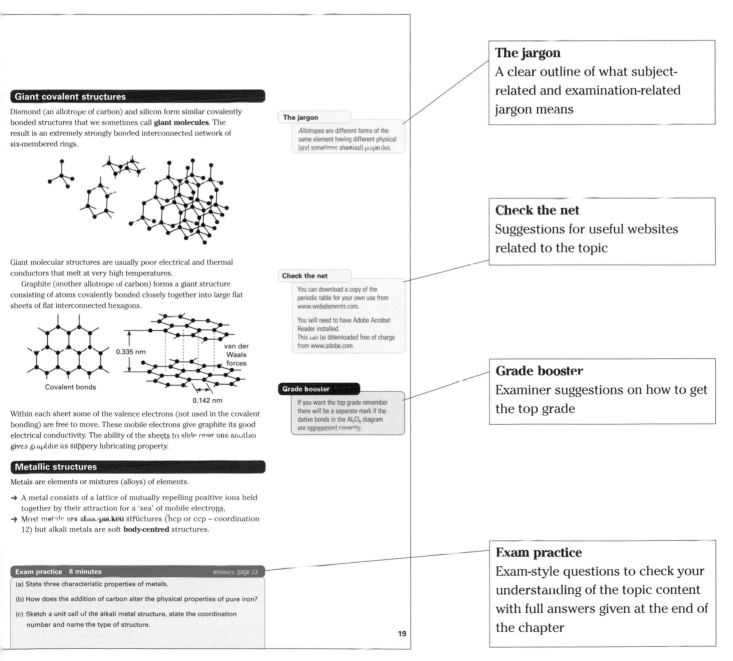

Giant covalent structures

Diamond (an allotrope of carbon) and silicon form similar covalently bonded structures that we sometimes call **giant molecules**. The result is an extremely strongly bonded interconnected network of six-membered rings.

Giant molecular structures are usually poor electrical and thermal conductors that melt at very high temperatures.

Graphite (another allotrope of carbon) forms a giant structure consisting of atoms covalently bonded closely together into large flat sheets of flat interconnected hexagons.

0.335 nm

van der Waals forces

Covalent bonds

0.142 nm

Within each sheet some of the valence electrons (not used in the covalent bonding) are free to move. These mobile electrons give graphite its good electrical conductivity. The ability of the sheets to slide over one another gives graphite its slippery lubricating property.

Metallic structures

Metals are elements or mixtures (alloys) of elements.

→ A metal consists of a lattice of mutually repelling positive ions held together by their attraction for a 'sea' of mobile electrons.
→ Most metals are close-packed structures (hcp or ccp – coordination 12) but alkali metals are soft **body-centred** structures.

The jargon

Allotropes are different forms of the same element having different physical (and sometimes chemical) properties.

The jargon
A clear outline of what subject-related and examination-related jargon means

Check the net

You can download a copy of the periodic table for your own use from www.webelements.com.

You will need to have Adobe Acrobat Reader installed.
This can be downloaded free of charge from www.adobe.com.

Check the net
Suggestions for useful websites related to the topic

Grade booster

If you want the top grade remember there will be a separate mark if the dative bonds in the Al_2Cl_6 diagram are represented correctly.

Grade booster
Examiner suggestions on how to get the top grade

Exam practice   8 minutes   answers: page 23

(a) State three characteristic properties of metals.

(b) How does the addition of carbon alter the physical properties of pure iron?

(c) Sketch a unit cell of the alkali metal structure, state the coordination number and name the type of structure.

Exam practice
Exam-style questions to check your understanding of the topic content with full answers given at the end of the chapter

19

Specification map

Edexcel		AQA		OCR		WJEC		CCEA	
AS	A2	AS	A2	AS	A2	AS	A2	AS	A2
○	●	○		○		○		○	
○		○		○		○		○	
○		○		○		○		○	
○		○		○		○		○	
○		○		○		○		○	
○		○		○		○		○	
○		○		○		○		○	
○		○		○		○		○	
○		○		○		○		○	
○		○		○		○		○	
○		○		○		○		○	
○		○		○		○		○	
	●		●		●		●		●
			●						
○	●	○	●	○	●	○	●	○	●
	●		●		●		●		●
	●		●		●		●		●
	●		●		●		●		●
	●		●		●		●		●
○	●	○	●	○	●	○	●	○	●
	●		●		●		●		●
○	●	○	●	○	●	○	●	○	●
	●		●		●		●		●
	●		●		●		●		●
	●		●		●		●		●
	●		●		●		●		●
	●		●		●		●		●
○	●	○	●	○	●	○	●	○	●
	●		●		●		●		●
	●		●		●		●		●
	●		●		●		●		●
	●		●		●		●		●
	●		●		●		●		●
○		○		○		○		○	
○		○		○		○		○	
○		○		○		○		○	
	●		●		●		●		●
	●		●		●		●		●
	●		●		●		●		●
	●		●		●		●		●
	●		●		●		●		●
	●		●		●		●		●
	●		●		●		●		●
	●		●		●		●		●
○		○		○		○		○	
○		○		○		○		○	
○	●	○	●	○	●	○	●	○	●
○		○		○		○		○	
○		○		○	●	○	●	○	
	●		●		●		●		●
	●		●		●		●		●
	●		●		●		●		●
	●		●		●		●		●
	●		●		●		●		●
	●		●		●		●		●
○		○		○		○		○	
○		○		○		○		○	
○		○		○		○		○	
○		○		○		○		○	
○		○		○		○		○	
○	●	○	●	○	●	○	●	○	●
○	●	○	●	○	●	○	●	○	●
○	●	○	●	○	●	○	●	○	●
○	●	○	●	○	●	○	●	○	●
○		○	●	○	●	○	●	○	●
	●		●		●		●		●
○	●		●		●		●		●
		○	●	○	●	○	●	○	●
AS	A2	AS	A2	AS	A2	AS	A2	AS	A2
○	●	○	●	○	●	○	●	○	●
○	●	○	●	○	●	○	●	○	●
○	●	○	●	○	●	○	●	○	●
○	●	○	●	○	●	○	●	○	●
○	●	○	●	○	●	○	●	○	●

Atoms, ions and molecules

You should already know that everything is made of atoms. For A-level, you will need to know what atoms are made of and how they form ions, simple molecules and giant structures. You will also need to understand how the properties of elements and compounds depend upon their structures and the forces that hold them together.

By the end of this chapter you will be able to

- Write and balance different types of chemical equation and work out empirical formulae using masses or percentage composition by mass

- Work out numbers of electrons, protons and neutrons using atomic and mass number and give electron configurations using s, p and d notation and the arrows in boxes method

- Use mass spectra to work out A_r values

- Use the Avogadro constant and do calculations on reacting masses, gas volumes and concentrations of solutions

- Interpret plots of 1st ionization energy against atomic number

- Draw dot and cross diagrams to show ionic and covalent bonding; draw examples and describe ionic lattices, simple molecules, giant covalent structures and metallic bonding. Relate the properties of a substance to the type of bonding it contains

- Describe the main types of intermolecular forces between simple molecules

- Work out shapes of molecules using VSEPR theory and work out whether molecules are polar

- Use $pV = nRT$ and the gas laws

Topic checklist

	Edexcel		AQA		OCR		WJEC		CCEA	
	AS	A2	AS	A2	AS	A2	AS	A2	AS	A2
Formulae and equations	O		O		O		O		O	
Atoms and isotopes	O		O		O		O		O	
Amount of substance and the mole	O		O		O		O		O	
Atomic structure	O		O		O		O		O	
Structure and bonding in elements	O		O		O		O		O	
Structure and bonding in covalent compounds	O		O		O		O		O	
Structure and bonding in ionic compounds	O		O		O		O		O	
Intermolecular forces	O		O		O		O		O	
Gases	O		O		O		O		O	

Tick each of the boxes below when you are satisfied that you have mastered the topic

Formulae and equations

Every element has a unique symbol. Chemists combine symbols to write formulae of chemicals and use formulae to write equations for chemical reactions. A formula can be shorthand for the name, composition and structure of a substance. A formula can represent an amount of substance and an equation can be the basis of chemical calculations involving masses, volumes, concentrations, etc.

Types of formulae

Empirical (simplest) formula

→ shows *ratios of the numbers* of each atom (or ion) in a compound
→ shows the *amounts of each element* in one formula mass

You should know how to use the composition by mass to find the empirical formula of a compound.

1 Divide each mass (or % mass) by the A_r of each element.
2 Calculate the simplest whole number ratios.

Example: An oxide of rubidium

element	% mass	$\div A_r$	=	mole ratio
Rb	84.2	$\div 85.5 = 0.99$		1
O	15.8	$\div 16.0 = 0.99$		1

The empirical formula is RbO.

Ionic formula

The ionic formula of a binary ionic compound is usually but not always the same as the empirical formula. The oxide of rubidium in the above example has a formula mass of 203 g mol⁻¹ and its ionic formula is $(Rb^+)_2O_2^{2-}$. Ionic compounds form crystals containing millions of ions arranged in a regular structure based on a **unit cell**.
Ionic formulae are related to:

→ the **radius ratios** of the ions in a unit cell
→ the **coordination numbers** of the ions in the crystal

If you know the formulae of the constituent ions you can deduce the formula of an ionic compound by making the charges balance.

Example: Magnesium bromide
Mg^{2+} is two positive and Br^- is one negative, so one magnesium cation needs two bromide anions to balance the charges: $Mg^{2+}(Br^-)_2$.

Molecular formula

This is for elements and compounds that form **covalent molecules**. Molecular formulae show the *numbers of atoms* in molecules. The noble gases are monatomic and ozone is triatomic but the other elemental gases like chlorine, nitrogen and oxygen are diatomic. You can use the **octet rule** to predict the molecular formula of many simple compounds of the s- and

Checkpoint 1

What is the empirical formula of a substance whose composition by mass is 50% sulphur and 50% oxygen?

The jargon

Binary compound is a combination of two elements only.

Checkpoint 2

What is the ionic formula of
(a) potassium iodide
(b) calcium sulphate
(c) lithium oxide
(d) barium hydroxide
(e) aluminium sulphate?
(Hint: Al is in group III of the periodic table!)

Checkpoint 3

What is the molecular formula of
(a) argon, (b) chlorine, (c) ozone?

Action point

Write an explanation of the octet rule in your own words.

p-block but *not* the d-block elements.

Example: Chlorine oxide

Chlorine is in group VII of the periodic table: $8 - 7 = 1$
Oxygen is in group VI of the periodic table: $8 - 6 = 2$

So one O atom needs two Cl atoms to balance: Cl_2O.

Structural formula

A structural formula shows the *arrangement* of atoms and groups in a molecule. Structural formulae are essential for organic compounds.

Checkpoint 4

Predict the molecular formula of the simplest compound formed by each of the following pairs of elements.
(a) chlorine and iodine
(b) sulphur and chlorine
(c) carbon and oxygen
(d) nitrogen and iodine

Types of equations

For **organic** reactions we often write unbalanced 'equations' showing only the structural formulae of the principal organic reactant(s) and product(s):

$$CH_3CH_2OH \longrightarrow CH_3CHO$$
$$\text{ethanol} \qquad \text{oxidized to} \qquad \text{ethanal}$$

For **inorganic** reactions we usually write balanced equations showing all the reactants and products.

Always write *balanced* equations for calculations.

Grade booster

Most marks are given for the correct formulae and balancing but it's a good idea to include the state symbols (s), (aq), (g) and (l) used properly.

Ordinary equation

The equation for the reaction of magnesium metal with dilute aqueous sulphuric acid is

$$Mg(s) + H_2SO_4(aq) \rightarrow MgSO_4(aq) + H_2(g)$$

Watch out!

Ionic equations must balance with respect to atoms and charges.

Ionic equation

The ionic equation for the reaction is

$$Mg(s) + 2H^+(aq) \rightarrow Mg^{2+}(aq) + H_2(g)$$

because the aqueous sulphate ion $SO_4^{2-}(aq)$ is a **spectator ion**.

Links

The atom economies for the reaction of magnesium with aqueous sulphuric acid are 98% magnesium sulphate and 2% hydrogen. See page 9.

Ion–electron half-equations

We can write the above ionic equation in the form of two ion–electron half-equations.

$$Mg(s) \rightarrow Mg^{2+}(aq) + 2e^- \qquad \text{oxidation – loss of electrons}$$
$$2H^+(aq) + 2e^- \rightarrow H_2(g) \qquad \text{reduction – gain of electrons}$$

You can use **oxidation numbers** to help balance equations for redox reactions.

Links

See page 72: redox reactions.

Exam practice (2 minutes)   answers: page 23

Write balanced ordinary and ionic equations for the reaction of

(a)   KOH(aq) with HNO_3(aq)

(b) $Ba(OH)_2$(aq) with HCl(aq)

(Hint: these are strong acids and bases.)

Watch out!

In ionic equations and ion–electron half-equations the symbols *and* the charges must balance!

Atoms and isotopes

Watch out!

Work in nuclear and theoretical physics suggests a much more complicated structure for the atomic nucleus but this is outside the scope of A-level and of little or no consequence to ordinary chemistry.

Experiments support the theory that atoms consist of a tiny nucleus of protons and neutrons surrounded by a large volume of space which, apart from the electrons, is empty. The negative charge of the electrons balances the positive charge of the protons. Most of the mass of an atom is in the nucleus. Isotopes of an element have different numbers of neutrons in their nuclei. Some isotopes are radioactive.

Electrons, protons and neutrons

The following typical diagram is not to scale because the nucleus is too small compared to the surrounding space occupied by the electrons.

Checkpoint 1

What is the name and symbol of the element with this atomic structure?

Name	Relative mass	Relative charge
● electron	1/1836	−1
● proton	1	+1
○ neutron	1	0

Checkpoint 2

What is the atomic number and mass number of the atom represented in this diagram?

nucleus (not to scale): 6 protons and 7 neutrons

The jargon

Z is used to number the element's place in the periodic table. Nowadays Z is often called the proton number.

→ **Atomic number** (Z) is the number of protons in the nucleus of an atom.
→ **Mass number** (A) is the number of protons and neutrons in a nucleus.

Isotopes

All the atoms of a particular element must have the *same number* of protons in the nucleus but the number of neutrons may be *different*.

→ Isotopes are atoms with the *same* atomic number (Z) but *different* mass numbers (A).

Checkpoint 3

How many neutrons are in the nucleus of each bromine isotope?

Bromine has two naturally occurring isotopes: $^{79}_{35}$Br and $^{81}_{35}$Br.

Radioactive isotopes

Some isotopes are unstable. Their nuclei spontaneously disintegrate and radiate *either* helium nuclei (alpha (α) particles) *or* electrons (beta (β) particles) but never both. Gamma (γ) rays (very high energy radiation) may also be given off. The **half-life** of a radioisotope is

Checkpoint 4

(a) How many protons and neutrons are in a helium nucleus?
(b) What happens to Z and A when a nucleus emits
   (i)   one α-particle
   (ii) one β-particle?

→ the time taken for its radioactivity to fall to half of its initial value
→ independent of the mass of the radioisotope, so the radioactive decay of an isotope is a **first order rate of reaction**
→ characteristic of each radioisotope and unaffected by catalysts or changes in temperature

Stable isotopes

The atoms of the majority of elements exist as a mixture of isotopes. Most stable nuclides have an even number of protons and/or neutrons in the nucleus. For atomic numbers up to 20, the ratio of the number of neutrons to the number of protons in stable nuclides is $1:1$.

Relative atomic mass A_r

The actual mass of an atom is so small (the heaviest is only about 4×10^{-25} kg) that we use relative atomic masses. We compare the mass of atoms to the mass of carbon atoms.

→ The **relative atomic mass** (A_r) is the *ratio* of the *average mass* per atom of the natural isotopic composition of an element *to one-twelfth of the mass of an atom of nuclide ^{12}C*.

Abundance of stable isotopes

The natural isotopic abundance of elements has been very accurately determined from mass spectra obtained from **mass spectrometers**. You could be asked to

→ read these abundances from a given mass spectrum
→ use them to calculate the relative atomic mass

Example: Mass spectrum of magnesium

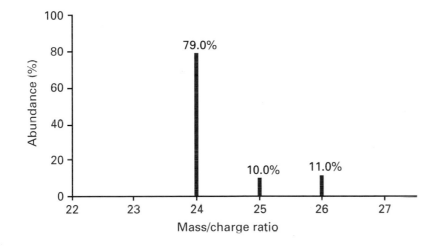

A_r(Mg) is $(79.0/100) \times 24 + (10.0/100) \times 25 + (11.0/100) \times 26 = 24.32$

The jargon

A *nuclide* is an atom of a specified mass number, A. All stable naturally occurring aluminium atoms exist as the nuclide $^{27}_{13}Al$.

Checkpoint 5

Why is the relative atomic mass of carbon itself 12.01 and not 12.00?

Links

See page 166: mass spectrometry.

Watch out!

A_r(Mg) is sometimes called the atomic weight of magnesium but don't forget that A_r stands for a ratio, so 24.32 is a pure number – no units!

Grade booster

Do not learn the A_r values. In an exam they are given to you to the first (or sometimes second) decimal place: see the periodic table on page 90

Watch out!

Candidates often lose marks by taking the Z value instead of the A_r value from the periodic table. Don't make this mistake if you want the top grade.

Grade booster

Candidates are often asked to explain the mass spectrum of chlorine. Chlorine contains Cl^{35} and Cl^{37} isotopes in the ratio 3:1 abundance.
In the Cl^+ region there are two peaks at $Z/m = 35$ and 37 in the ratio of 3:1.
In the molecular ion (Cl_2^+) region of the spectrum there are three peaks at $Z/m = 70$, 72, and 74 in the ratio of 9:6:1.

Exam practice (5 minutes)   answers: page 23

35 and 37 are the mass numbers of the two stable naturally occurring isotopes of chlorine.

(a) State for each nuclide (i) the number of protons in the nucleus and
    (ii) the number of neutrons in the nucleus.

(b) Calculate the relative abundance of the two isotopes of chlorine (refer to the periodic table on page 90).

Amount of substance and the mole

When we read an equation we can see from the numbers and formulae how much of each reactant and product there is. The chemist's unit of 'how much' is the mole. You need to understand how it is defined and how we use it in calculations.

Checkpoint 1

Use the Periodic Table on page 90 to determine
(a) the mass in grams of
(i) 1 mol of hydrogen atoms
(ii) 0.5 mol of nitrogen molecules, N_2
(iii) 1 mol of ozone molecules, O_3
(b) The molar mass of nitric acid, HNO_3

The jargon

$n = m/M$
where n is the amount of substance in moles, m is the mass of the substance in grams and M is the molar mass in grams per mole.

Grade booster

You do not need to remember values for molar gas volumes.

The jargon

The symbol for the Avogadro constant is an italic capital L.

Don't forget!

Since L is a number per mole, the units of L will be just mol^{-1}.

Grade booster

You are always given the value of L to two (or three) significant figures. You do not need to remember it.

What is a mole?

One mole is the mass, in grams, of a substance as indicated by its formula. How can you find the mass of a mole? Look up the formula and **molar mass** in data books: e.g.

Formula	Na	Cl^-	H_2O	Na^+Cl^-
Molar mass/g mol^{-1}	23.0	35.5	18.02	58.5

You can also use the formula and the A_r values of the constituent elements to calculate the **relative formula mass**. Here are the steps:

1 The formula of water is H_2O.
2 $A_r(H) = 1.01$ and $A_r(O) = 16.0$.
3 The relative formula mass is $(2 \times 1.01 + 16.0) = 18.02$.
4 So the mass of one mole of H_2O would be 18.02 g.

What is an exact definition of a mole?

→ One mole (mol) is the amount of substance (n) that contains as many entities (atoms, molecules, electrons or other particles) specified by a formula as there are carbon atoms in 0.012 kg of carbon nuclide ^{12}C.

Using the Avogadro constant

Example 1:   How many molecules are there in 18.02 g of water?
In 0.012 kg of carbon nuclide ^{12}C there are about 6.02×10^{23} atoms.
So in 18.02 g of water there will be about 6.02×10^{23} H_2O molecules.

Example 2:   How many molecules are there in 45.05 g of water?
A molar mass of H_2O is 18.02 g mol^{-1}. So, the amount of H_2O in 45.05 g water is $45.02/18.02 = 2.5$ mol, and therefore the number of H_2O molecules is $(6.02 \times 10^{23}\ mol^{-1}) \times (2.5\ mol) = 1.505 \times 10^{24}$.

How do we find the number of entities in any given amount of substance?

number of entities = 6.02×10^{23} × amount of substance
number of entities = L × amount of substance

→ The Avogadro constant is the proportionality constant connecting the number of entities (specified by a formula) to the amount of substance (expressed in moles).

Molar volume

This is the volume of one mole of substance under *specified conditions of temperature and pressure*.

In 1811 the Italian chemist Amadeo **Avogadro** stated this important principle:

→ Equal volumes of gases at the *same temperature and pressure* contain the same number of molecules.

One mole of any gas occupies approximately 24 dm^3 at room temperature and 22.4 dm^3 at standard temperature and pressure (s.t.p.). According to the rule proposed by the French chemist Louis J. **Gay-Lussac** in 1809

→ when measured at the same temperature and pressure the volumes of gaseous reactants and products of a reaction will be in a simple ratio.

Example: $\quad H_2(g) \quad + \quad Cl_2(g) \quad \rightarrow \quad 2HCl(g)$
$\qquad\qquad$ 1 mol or 24 dm³ $\quad$ 1 mol or 24 dm³ $\quad$ 2 mol or 48 dm³

so the gases are in the ratio 1 : 1 : 2.

The jargon

Standard temperature is 273.15 K.
Standard pressure is 101.325 kPa.

Concentration

When 6.30 g of nitric acid is dissolved in water to produce exactly 1.00 dm³ of solution, the resulting aqueous acid should be described as 'aqueous nitric, $HNO_3(aq)$, of concentration 0.100 mol dm⁻³'.

→ Concentration is the amount of substance per unit volume of *solution* (not solvent) expressed in mol dm⁻³.

Checkpoint 2

For 16.0 g of oxygen at s.t.p. what would be
(a) the volume of gas and
(b) the number of oxygen molecules, O_2?

Chemical calculations

You need to be able to use amounts of substances, molar gas volumes and concentrations of solutions to do calculations and solve problems based on balanced chemical equations. Here are two examples.

1. What volume of hydrogen at s.t.p. would be produced by reacting completely 0.486 g of magnesium with excess aqueous sulphuric acid? The equation for the reaction is

$$Mg(s) + H_2SO_4(aq) \rightarrow MgSO_4(aq) + H_2(g)$$

1 mol = 24.3 g would produce 1 mol = 22.4 dm⁻³ hydrogen so 0.486 g magnesium would produce 22.4 × 0.486/24.3 = 0.448 dm⁻³.

2. What is the *minimum* volume of hydrochloric acid of concentration 0.01 mol dm⁻³ needed to react completely with 0.243 g of magnesium? The equation is $Mg(s) + 2HCl(aq) \rightarrow MgCl_2(aq) + H_2(g)$, so 0.243 g = 0.01 mol Mg needs 0.02 mol HCl(aq) or 2 dm³ of solution.

Grade booster

You do not need to remember values for molar gas volumes.

Grade booster

The arithmetic is kept as simple as possible because your chemistry is being tested more than your maths.

Atom economy

The atom economy is an indication of the efficiency of a reaction and is defined as $\left[\dfrac{\text{formula mass of a product}}{\text{total formula mass of reactants}}\right] \times 100\ \%$

For the above reaction of magnesium (r.a.m. = 24.3) with sulphuric acid (r.m.m. = 98.12), the total formula mass of reactants is 2.3 + 98.12 = 122.42. So the atom economy for $MgSO_4$ is 100 × (120.4/122.42) = 98.35% and the atom economy for H_2 is 100 − 98.35 = 1.65%.
NB. If **all** the hydrogen in the sulphuric acid is released as $H_2(g)$ then the percentage yield of hydrogen would be 100%

Watch out!

Do not confuse atom economy with percentage yield.

Action point

Compare the atom economy and the percentage yield of ammonia in the Haber process (see page 119).

Exam practice (6 minutes) $\qquad\qquad$ answers: page 23

Calcium carbonate, $CaCO_3$, is obtained from limestone rock by open quarrying for large-scale industrial use in, for example, the production of calcium oxide, CaO, by roasting in high-temperature rotary furnaces.
(a) Write an equation for the decomposition of calcium carbonate.
(b) Calculate (i) the mass of calcium oxide and (ii) the volume of carbon dioxide, at 25 °C and 1 atm, that could be produced from 1 000 kg of calcium carbonate.
(c) Comment upon two environmental concerns regarding the industrial use of limestone.

Watch out!

For the top grade remember to balance the equation and include the state symbols (s) and (g). Look up the appropriate value for the molar gas volume. You should know some formulae so some examiners may not give you the formula of calcium carbonate or calcium oxide!

Atomic structure

This section is all about the way electrons are arranged in shells, subshells and orbitals of atoms and ions. You need to understand how to work out these electron arrangements from the patterns in the ionization energies of atoms; but first, what is an ion and what is ionization energy?

The jargon

Atomic orbitals are mathematical descriptions of probability distributions of electrons in the space surrounding a nucleus.

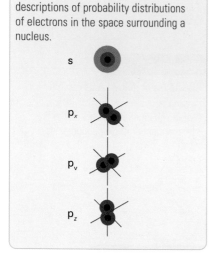

Formation of ions

$$Na \rightarrow Na^+ + e^- \text{ and } Mg \rightarrow Mg^{2+} + 2e^-, \text{ etc.}$$
$$Cl + e^- \rightarrow Cl^- \text{ and } O + 2e^- \rightarrow O^{2-}.$$

→ Metal atoms lose electrons to form cations.
→ Non-metal atoms gain electrons to form anions.

Ionization energies

If we supply enough energy (electrical or thermal) to any gaseous atoms they can lose electrons and become ionized.

→ The **molar first ionization**, E_{m1}, of an element is the energy required to remove one mole of electrons from one mole of its gaseous atoms: $X(g) \rightarrow X^+(g) + e^-$.
→ The **molar second ionization**, E_{m2}, of an element is the energy required to remove one mole of electrons from one mole of its gaseous unipositive ions: $X^+(g) \rightarrow X^{2+}(g) + e^-$.

Watch out!

We plot the *logarithm*, E_{mj}, of the ionization energies, not the E_{mj} values.

Make sure you can relate the pattern of *successive* ionization energies for an element to the principal electron energy levels in its atoms.

Checkpoint 1

What is the atomic number and name of the element this pattern refers to?

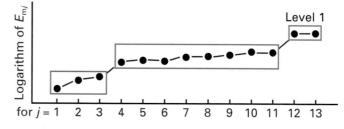

You should also be able to relate the pattern of molar first ionization energies for the first 20 elements to the arrangement of electron energies into subsets.

Watch out!

Here we plot the value of the first ionization, E_{m1}, for each of the elements.

Checkpoint 2

(a) What elements are numbers 2, 10 and 18 at the peaks?
(b) On the chart mark in blue the E_{m1} values for the alkali metals.
(c) On the chart mark in red the E_{m1} values for the halogens.

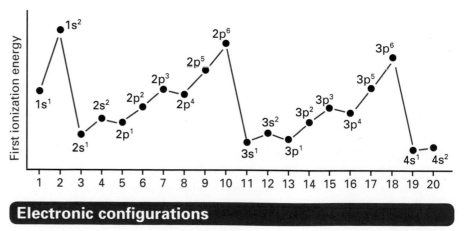

Electronic configurations

The energy levels of electrons in atoms are arranged into sets which are numbered 1, 2, 3, etc. and called shells. This arrangement is called the electronic configuration which, for potassium and chlorine for example, we write K 2.8.8.1 and Cl 2.8.8.7. The last number in these sequences (1 and 7) refers to the so-called '**valence electrons** in the **outer shell**'. The numbers 2.8.8 refer to the so-called '**inner shell electrons**'.

→ Valence electrons have the lowest ionization energies and are involved in bonding.
→ Inner shell electrons have higher ionization energies and partially shield the outer (valence) shell electrons from the nuclear charge.
→ Atoms of elements in the same group of the periodic table have the same number of valence electrons in their outer shell.
→ The maximum numbers of electrons at each of the first five levels of energy are 2.8.18.32.32.

The jargon

The numbers 1, 2, 3, 4, 5, 6, 7 are called *principal quantum numbers*. The electronic configurations given in data books are for the atoms in their ground states.

Subshells and orbitals

The principal energy levels of electrons in atoms can be arranged into subsets which are labelled s, p, d, f and called **subshells**.

→ Electrons in the same subshell do not shield each other very much from the attraction of the nucleus.

The subshell energy levels can be arranged into subsets called orbitals. The s-subshell has one orbital, the p-subshell has three, the d-subshell has five and the f-subshell has seven orbitals.

→ No more than two electrons may be in the same orbital.

The jargon

The letters s, p, d and f stem from the words *sharp*, *principal*, *diffuse* and *fundamental* used to describe the lines in spectra.

The jargon

We say that two electrons in the same orbital have *opposite spins*.

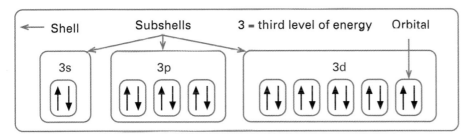

Checkpoint 3

What is the electronic configuration of
(a) Li and Na
(b) F and Br
(c) He, Ne and Ar?

Building up electronic configurations

Use this plan to work out ground state electronic configurations:

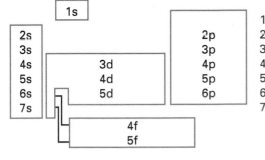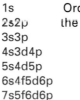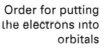

	Order for putting
1s	the electrons into
2s2p	orbitals
3s3p	
4s3d4p	
5s4d5p	
6s4f5d6p	
7s5f6d6p	

Watch out!

When you have worked out the number of electrons in each subshell you must write adjacent to each other the subshells with the same principal quantum number. For example:

$1s^22s^22p^63s^23p^64s^23d^3$ is wrong
$1s^22s^22p^63s^23p^63d^34s^2$ is right

Bear this in mind if you want the top grade.

Exam practice (3 minutes)   answers: page 23

Write the ground state electronic configurations of atoms with atomic numbers 17, 24 and 33.

Structure and bonding in elements

So what holds things together? Here you will see why nitrogen is a gas with hardly any attraction between one molecule and another, but carbon, in the form of diamond, its next-door-neighbour in the periodic table, is a solid and one of the hardest substances known. You will also see why metals conduct and why non-metals don't. Watch out for the exception to this rule!

Noble gases and van der Waals forces

The jargon

van der Waals forces are weak instantaneous dipole–dipole attractions operating between atoms in all substances.

The noble gas molecules are **monatomic**. Their electron shells are full so the atoms do not combine with each other. However, at very low temperature and high pressure the noble gases condense to liquids in which the atoms are held together by very weak van der Waals forces.

→ The more electrons there are in an atom the more van der Waals forces there will be.

Checkpoint 1

State and explain the trend in b.p. of the noble gases with increasing atomic number from helium to xenon.

Chemists have not yet made any compounds of helium, neon or argon. So the electronic configurations of their atoms and those of the other noble gases are considered to be particularly stable arrangements of electrons. Many (but not all) chemically reactive elements combine so that the electrons in their atoms rearrange to match the electronic configurations of noble gases.

→ Atoms have a tendency to achieve a noble gas electronic configuration.

Simple molecules and the covalent bond

→ A covalent bond is the result of two nuclei *sharing* electrons.
→ A covalent bond forms because the attraction of the nuclei for the electrons is greater than the repulsions between the nuclei and between the electrons of the two atoms.

Watch out!

We represent electrons by dots and crosses only as a convenience to show their origin and help us keep count of the number of valence electrons in each atom. And we usually omit the inner shell electrons.

Hydrogen, the halogens, oxygen and nitrogen molecules are diatomic. We can represent these molecules by 'dot and cross' Venn diagrams:

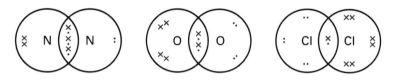

where the overlap shows the sharing of electrons.

The jargon

The — represents a covalent bond of two shared electrons. The **:** represents a lone pair of non-bonded electrons.

$$:N\equiv N: \qquad \ddot{O}=\ddot{O} \qquad :\ddot{C}l-\ddot{C}l:$$

Triple bond   Double bond   Single bond

Sulphur and white phosphorus form P_4 and S_8 molecules.

Checkpoint 2

Draw a dot and cross Venn diagram for a molecule of
(a) neon
(b) hydrogen

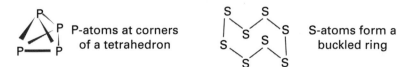

P-atoms at corners of a tetrahedron   S-atoms form a buckled ring

Simple molecular structures are usually poor conductors that melt and boil at low temperatures, consuming little heat in the process.

Giant covalent structures

Diamond (an allotrope of carbon) and silicon form similar covalently bonded structures that we sometimes call **giant molecules**. The result is an extremely strongly bonded interconnected network of six-membered rings.

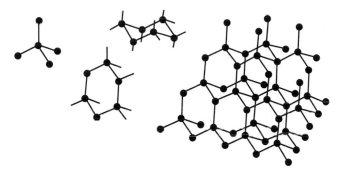

The jargon

Allotropes are different forms of the same element having different physical (and sometimes chemical) properties.

Grade booster

If you draw a diagram, keep it simple, e.g. for diamond, just show the tetrahedral bonding (five atoms) and indicate an extended structure with two or three more atoms. Do **NOT** try to reproduce textbook diagrams.

Giant molecular structures are usually poor electrical and thermal conductors that melt at very high temperatures.

Graphite (another allotrope of carbon) forms a giant structure consisting of atoms covalently bonded closely together into large flat sheets of flat interconnected hexagons.

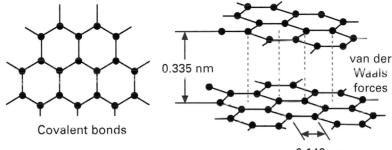

Covalent bonds

0.335 nm

van der Waals forces

0.142 nm

Watch out!

An extremely pure graphite crystal is hard and its electrical conductivity is low in one direction.

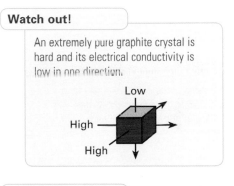

Low

High

High

Within each sheet some of the valence electrons (not used in the covalent bonding) are free to move. These mobile electrons give graphite its good electrical conductivity. The ability of the sheets to slide over one another gives graphite its slippery lubricating property.

The jargon

The mobile electrons in graphite and in metals are called *delocalized electrons*. *Carbon nanotubes* are 10^4 times thinner than a human hair and are like rolled up graphite layers of carbon hexagons.

Metallic structures

Metals are elements or mixtures (alloys) of elements.

→ A metal consists of a lattice of mutually repelling positive ions held together by their attraction for a 'sea' of mobile electrons.
→ Most metals are **close-packed** structures (hcp or ccp – coordination 12) but alkali metals are soft **body-centred** structures.

Links

See page 175: nanochemistry.

The jargon

An *alloy* is a solid solution of two or more metals. Sometimes a non-metal forms part of an alloy.

Exam practice (8 minutes)   answers: page 23

(a) State three characteristic properties of metals.

(b) How does the addition of carbon alter the physical properties of pure iron?

(c) Sketch a unit cell of the alkali metal structure, state the coordination number and name the type of structure.

Structure and bonding in covalent compounds

The bonding in binary non-metal compounds is covalent but bond polarization may produce some ionic character. You can use the valence shell electron pair repulsion (VSEPR) theory to predict the shapes of simple covalent molecules.

Simple covalent molecules

A covalent bond exists between two atoms when they share electrons.

→ **Bond length** is the average distance between the nuclei of the two atoms.
→ **Bond energy** is the energy needed to separate completely the atoms in the molecules of one mole of compound.

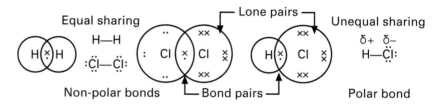

Equal sharing | Lone pairs | Unequal sharing

H—H   :Cl—Cl: $\delta+$ $\delta-$ H—Cl:

Non-polar bonds — Bond pairs — Polar bond

→ A dative (or coordinate) bond is a covalent bond in which one of the atoms has supplied *both* electrons being shared.

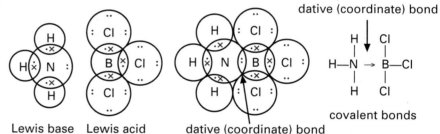

dative (coordinate) bond

$$H—N \rightarrow B—Cl$$

covalent bonds

Lewis base   Lewis acid   dative (coordinate) bond

Polar bonds

In a diatomic hydrogen (or chlorine) molecule the two identical nuclei must attract the bonded electron pair of electrons equally strongly.
In a molecule of HCl the chlorine nucleus (albeit shielded by two inner shells of electrons) attracts the electrons more strongly than does the hydrogen nucleus. This makes the H-atom *slightly* positive and the Cl-atom *slightly* negative. Consequently, the H–Cl bond is **polar** and the HCl molecule has a **dipole moment**.

Electronegativity

The **Pauling electronegativity index** (Np) is a measure of how strongly an atom in a compound attracts electrons in a bond.

Trends in electronegativity index values in the periodic table:

Increasing electronegativity

H (2.1)	B (2.0)	C (2.5)	N (3.0)	O (3.5)	F (4.0)
		Si (1.8)	P (2.1)	S (2.5)	Cl (3.0)
			As (2.0)	Se (2.4)	Br (2.8)
				Te (2.1)	I (2.5)

The greater the difference in the electronegativities of two atoms, the more polar is the covalent bond between the two atoms.

Don't forget!

For any two given atoms the bond gets shorter and stronger as it changes from — to = to ≡.

Watch out!

Dots and crosses help to show which valence electrons come from which atom *but* once a bond is formed the electrons are indistinguishable.

Checkpoint 1

Draw a dot and cross diagram for a molecule of (a) methane, (b) carbon dioxide and (c) iodine chloride.

The jargon

A *Lewis acid* is a molecule or ion that can accept a pair of electrons to form a covalent bond. What is a Lewis base? See page 63 for the meaning of a Brønsted–Lowry acid and base.

Grade booster

You don't need to remember Np values but you do need to learn these really important trends if you want the top grade.

Checkpoint 2

Which halogen is most electronegative and which hydrogen halide is the least polar?

Polar molecules

Oxygen is more electronegative than carbon and more electronegative than hydrogen. Consequently the bonds between oxygen and these elements are polarized so the O-atom is slightly negative.

Why is water a polar molecule while CO_2 is not?

In carbon dioxide the two CO bonds are equally polar but their dipoles act in opposite directions and cancel each other out.

In water the dipoles of the two HO bonds act in similar directions and reinforce each other.

answers: page 24

Shapes of molecules

You should be able to predict the approximate angles between bonds and the shape of simple molecules using rules based on the **VSEPR** theory.

1   Write formulae to show all electron (− bonded and : lone) pairs.

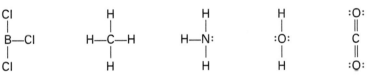

2   Assume the electron pairs (bonded and lone) move equally as far apart as possible from each other but treat double bonds as single pairs.

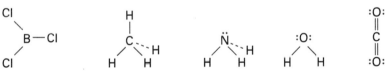

Trigonal 120°   Tetrahedral 109.5°   Pyradmidal   Non-linear   Linear 180°

3   Adjust any bond angles affected by the following rule for repulsion between bonded (bp) and non-bonded (nbp) electron pairs:

nbp.nbp repulsion > nbp.bp repulsion > bp.bp repulsion

Pyramidal 107°   Non-linear 105°

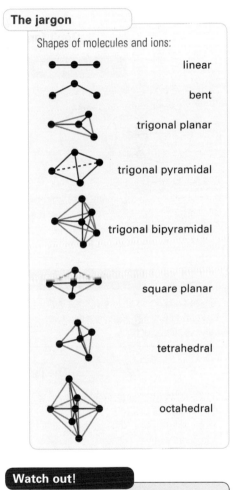
Exam practice (10 minutes)

(a) Draw dot and cross diagrams to represent the following covalent molecules: (i) H_2S, (ii) NH_3, (iii) PCl_5, (iv) Al_2Cl_6.

(b) For each of the molecules in (a) predict and draw the shape of the molecule and indicate its likely polarity, if any.

Structure and bonding in ionic compounds

The jargon

Anion polarization is distortion of the shape of a polarizable anion by a polarizing cation. *Cation polarizing power* is the ability to distort the shape of anions.

Checkpoint 1

What combination of cation and anion would give a binary ionic compound with the least covalent character?

Watch out!

Do *not* overlap the circles like this:

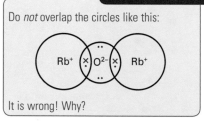

It is wrong! Why?

Checkpoint 2

Draw a dot and cross diagram for the formation of calcium fluoride.

Grade booster

You can omit inner shell electrons but you will lose marks if you leave out any valence shell electrons.

The jargon

Simple cubic: the *same ion* at each of the eight corners of a cube.
Double simple cubic: the cubes interlock.
Face-centred cubic: the *same ion* at each of the eight corners of a cube *and* in the centre of each face.

Watch out!

Strictly speaking NaCl is a double (or interlocking) face-centred cubic structure.

Grade booster

Learn these two structures and coordination numbers. They are popular with examiners.

Checkpoint 3

Explain why these two chlorides are
(a) hard crystalline solids with high melting points and
(b) electrical non-conductors when solid but conductors when molten.

The bonding in binary metal-non-metal compounds is ionic but cations may polarize anions to produce some covalent character. You can use the octet rule and electronic configurations to predict the formula of simple compounds.

Simple ionic bonding

Ionic bonding is the result of electrons being transferred from one atom to another and the ions packing together into a crystal **lattice**.

Predicting simple ionic formulae

Example: The formula of rubidium oxide

The atomic number of rubidium is 37 and of oxygen is 8. Their ground state electronic configurations are

$$\text{Rb } 2.8.18.8.1 \text{ and O } 2.6}$$

Predict Rb to lose one electron and O to gain two electrons:

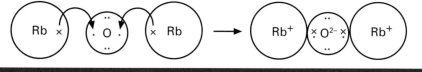

Ionic crystal structures

You should be able to

→ draw diagrams and describe the unit cells of CsCl and NaCl
→ work out the number of ions in each unit cell
→ deduce the empirical formula from the ratio of the coordination numbers of each ion

The **coordination number** of an ion in a crystal of an ionic compound is the number of nearest, equidistant, oppositely charged ions.

Caesium chloride
Double simple cubic

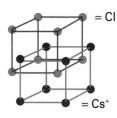

= Cl
= Cs⁺

Coordination no. $Cs^+ = 8$
Coordination no. $Cl^- = 8$
Ratio of $Cs^+ : Cl^- = 1 : 1$
Hence formula is CsCl

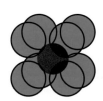

Sodium chloride
Face-centred cubic

What is the coordination no. $Na^+ =$ ____
What is the coordination no. $Cl^- =$ ____
What is the ratio of $Na^+ : Cl^- =$ __ : __
Hence formula is NaCl

= Cl⁻
= Na⁺

Crystal systems

There are seven different crystal systems: cubic, hexagonal, monoclinic, orthorhombic, rhombohedral, tetragonal and triclinic. The type of crystal structure adopted by an ionic compound is governed by the relative numbers, shapes and sizes of the cations and anions.

→ **Radius ratio** is the radius of the smaller ion divided by the radius of the larger ion.

Hydrated crystal structures

Cations attract the negative (oxygen) end of water molecules and anions attract the positive (hydrogen) end of water molecules.

→ Many ionic compounds form hydrated crystals in which the water molecules are held in the lattice by **ion–dipole forces**.
→ Hydrated ionic compounds melt at much lower temperatures than the corresponding anhydrous compounds.
→ Less heat is required to melt hydrated ionic compounds which often decompose by loss of water.

Complex cations

Transitional metal cations form complex ions in which non-metal ligands are attached to the metal ion by **dative bonds**.

Polyatomic non-metal ions

Many molecules may form anions (or cations) by acting as acids (or bases) and losing (or gaining) protons.

You should be able to draw dot and cross diagrams and predict the shape of some simple anions and cations.

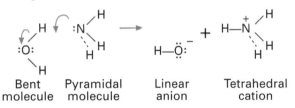

| Bent molecule | Pyramidal molecule | Linear anion | Tetrahedral cation |

Delocalized bonding

When you use the **octet** and **VSEPR** theories to predict the shapes of certain molecules or ions and you find that you can write more than one possible arrangement of the bonds, the actual structures are probably delocalized.

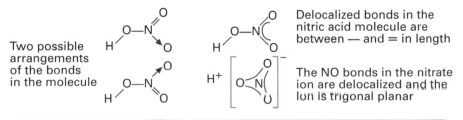

Two possible arrangements of the bonds in the molecule

Delocalized bonds in the nitric acid molecule are between — and = in length

The NO bonds in the nitrate ion are delocalized and the ion is trigonal planar

Checkpoint 4

(a) Use the following ionic radii (in nm) to calculate the radius ratios in caesium chloride and sodium chloride:
Cs$^+$ = 0.170; Na$^+$ = 0.102; Cl$^-$ = 0.180
(b) What crystal system is adopted by (i) caesium chloride, (ii) sodium chloride?

The jargon

Ion–dipole interaction is the attraction between an ion and a polar molecule. *Anhydrous* means no water of crystallization.

The jargon

A *ligand* is a molecule or anion with one or more atoms having a lone pair of electrons for donation to form a dative bond.

Links

See page 127: ligands.

The jargon

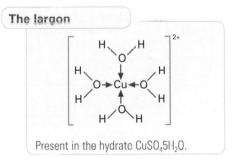

Present in the hydrate CuSO$_4$5H$_2$O.

Exam practice (10 minutes)   answers: page 24

(a) Draw a dot and cross diagram for (i) sodium hydride, (ii) ammonium chloride, (iii) the oxonium ion.

(b) Suggest, in terms of bonding, why the melting point of aluminium fluoride is much higher than the melting point of aluminium chloride.

(c) Deduce the structure and shape of the carbonate ion.

Intermolecular forces

We know that a molecule is held together by strong covalent bonds but what holds one molecule to another? Why are they attracted at all? Why is water a liquid? Why does the boiling point of the halogens increase down the group? Read on!

van der Waals forces

van der Waals forces are weak *short-range* forces between atoms and molecules arising from the attraction between dipoles. These forces arise from the movement of the electrons in relation to the nuclei which produces *weak instantaneous dipoles* that attract one another.

→ The more electrons there are in an atom, the more instantaneous dipole–dipole attractions there will be.
→ van der Waals forces account for the increase in boiling point with increasing molar mass of **noble gases** and **halogens**.

Checkpoint 1

Suggest why the b.p. of iodine chloride (ICl) is higher than the b.p. of bromine (Br_2).

Dipole–dipole attractions

Dipole–dipole forces are the attractions between the positive end of one *permanently* polar molecule and the negative end of another *permanently* polar molecule.

Ion–dipole interactions

Many ionic compounds dissolve well in water and other polar solvents because the energy released by the formation of the ion–dipole interactions compensates for the **lattice energy** consumed in overcoming the electrostatic forces between the cations and anions.
We picture the process as follows:

The jargon

Hydration energy is the enthalpy change for the formation of one mole of the aqueous ions from ions in the gaseous state.

Don't forget

Enthalpy changes of solution are usually small and may be exothermic or endothermic.

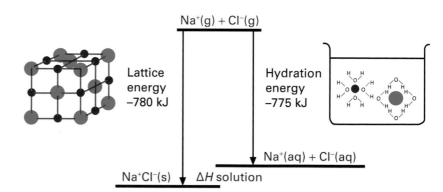

Checkpoint 2

(a) What is the molar enthalpy change of solution of sodium chloride?
(b) Does salt dissolve in water exothermically or endothermically?

1 Cations and anions attract polar water molecules.
2 Water molecules get between the ions and lessen their attractions.
3 Water molecules completely surround cations and anions.
4 Hydrated ions leave the crystal and mix with the water.

Grade booster

Top grade candidates will be able to explain the dissolving of ionic compounds in water in terms of entropy changes (see page 84) – the disorder of the dissolved ions is greater than the order of the ionic lattice and the hydration layer around the dissolved ions.

Hydrogen bonding

A hydrogen bond (···) is a weak bond between a very electronegative atom (X = N, O or F) and a hydrogen atom bonded to a very electronegative atom (Y = N, O or F). Thus —X···H—Y.

→ Hydrogen bonding is stronger than van der Waals forces and permanent dipole–dipole attractions but weaker than covalent bonding.

You should be able to explain the exceptionally high boiling points of ammonia, water and hydrogen fluoride (compared to the values for the other hydrides in each group) by hydrogen bonding.

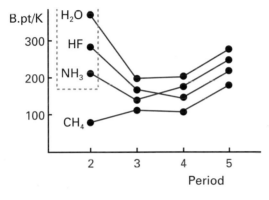

Hydrogen bonding is a major factor in determining the structure and properties of water, hydrated salts, carbohydrates and proteins.

Dimerization of acids

Hydrofluoric acid is atypical in being a weak acid (pK_a = 3.3) because it **dimerizes** in water, due to hydrogen bonding. This decreases the number of protons donated to the water molecules:

$$H_2O(l) + H—F···H—F(aq) \rightarrow H_3O^+(aq) + [F···H—F]^-(aq)$$

In pure crystals or anhydrous liquids, carboxylic acids form **dimers**, due to hydrogen bonding.

Example:

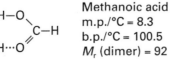

Methanoic acid
m.p./°C = 8.3
b.p./°C = 100.5
M_r (dimer) = 92

Exam practice (10 minutes)   answers: pages 24–5

(a) Explain each of the following.

  (i) The boiling point of hydrogen fluoride, HF, is the highest of all the hydrogen halides.

  (ii) In non-aqueous media, ethanoic acid exists as dimeric molecules ($CH_3CO_2H)_2$.

  (iii) Water in the solid state is less dense than water in the liquid state.

(b) (i) State the type of ionic crystal structure and the coordination number of the cation in caesium chloride.

  (ii) Explain why the answer in (b) (i) would be different for sodium chloride.

Gases

This section is about gases and the laws that describe their behaviour. You need to understand the ideal gas equation and know how to use it.

The jargon

Kinetic comes from the Greek kinetikos, to move.

Kinetic molecular theory of gases

We explain the physical properties of gases by picturing tiny molecules in constant rapid and random motion colliding with themselves and with the walls of their container. In this theory for an **ideal gas** we make the following important assumptions.

→ The volume of the molecules themselves is negligible compared with the volume of the containing vessel they occupy.
→ The molecules are not attracted to each other or to the walls of the container.
→ All collisions are perfectly elastic and so the total kinetic energy of all the molecules is constant.
→ The mean kinetic energy of the molecules is directly proportional to the Kelvin temperature of the gas.
→ The collision of the molecules with the walls of their container is responsible for the observable gas pressure (force per unit area).

Watch out!

An Ideal or perfect gas obeys the experimental laws of Boyle and Charles and the equation $pV = nRT$ but real gases deviate from this ideality especially at high pressure and low temperature.

The gas laws

→ Boyle's law: pV = a constant

For a given mass of gas, at constant temperature, the volume (V) is inversely proportional to the pressure (p).

→ Charles' law: V/T = a constant

For a given mass of gas at constant pressure, the volume (V) is directly proportional to the Kelvin temperature (T).

→ Avogadro's law: $V \propto n$

For a gas at constant temperature and pressure, the volume (V) is proportional to the number of molecules (n).

Checkpoint

(a) Sketch the graph of volume against pressure for a given mass of ideal gas at constant temperature.
(b) Sketch the graph of volume against Kelvin temperature for a given mass of ideal gas at constant pressure.

The general gas equation

The three laws above can be combined into one equation.

→ ideal gas equation: $pV = nRT$
→ n (number of moles) = m/M

m is the mass of gas in grams and M is its molar mass.

The jargon

The value of the gas constant R is usually given as 8.314 J K^{-1} mol^{-1}. To be consistent with the units we need to use volume (V) in m^3, pressure (p) in N m^{-2} (= Pa) and temperature (T) in K. The units of molar mass (M) are g mol^{-1}.

Using the ideal gas equation

If we can measure the volume of a known mass of a compound at a known temperature and pressure, then we can determine the molar mass of the compound.

If the gas dissociates, then the value of M_r (the relative molecular mass) can tell us the degree of dissociation α at the given temperature and pressure. This in turn can lead to a value for K_p, the equilibrium constant for the dissociation.

Don't forget!

When you make measurements on a gas or vapour and do your calculations using $pV = nRT$ remember that you are assuming ideal behaviour.

Example: Calculate the molar mass of a volatile organic liquid X given that the volume of 0.567 1 g of its vapour is 145.2 cm^3 at 100 °C and 101.3 kPa.

The general gas equation $pV = (m/M)RT$ can be rewritten in the form

$$M = mRT/pV$$

Substituting for $m = 0.567$ 1 g, $V = 1.452 \times 10^{-4}$ m^3, $p = 101\,300$ Pa, $T = 373$ K gives

$$M = (0.567\ 1 \times 8.134 \times 373)/(101\ 300 \times 1.452 \times 10^{-4})$$
$$= 119.5 \text{ g mol}^{-1}$$

Grade booster

You will always be given the gas constant value, usually as $R = 8.314$ J mol^{-1} K^{-1}. Notice the units in J (= joules), *not* kJ (= kilojoules). If you want the top grade be sure to use the correct units for P, V and T.

Dalton's law of partial pressures

Make sure you understand this law. You often need to use partial pressure when calculating equilibrium constants.

Links

See pages 52–3: equilibria.

→ In a mixture of two or more gases, the partial pressure of each gaseous component is the pressure which that gas would exert if it alone occupied the volume taken up by the mixture of gases.
→ The total pressure (P) of a gaseous mixture equals the sum of the individual partial pressures (P) of the components: $P = \sum p$.
→ The partial pressure of any component in a mixture of gases is given by $p = (n_c/n_t) \times P$.

The jargon

n_c is the number of moles of the component gas
n_t is the total number of moles of gas in the mixture
n_c/n_t is called the mole fraction
The mathematical symbol, Σ, means 'the sum of'.

Real gases

At very high pressure and low temperature the volume of the molecules themselves becomes significant. Furthermore, the experimental fact that all gases can be liquefied under these conditions shows that there must be forces of attraction between molecules.

Action point

Make a list or table of the names and relative strengths of the bonding and attractive forces holding atoms and molecules together. This will help you get the top grade

→ All gases deviate considerably from the ideal behaviour at high pressures and low temperatures.

van der Waals' equation

J. C. van der Waals, a Dutch chemist, proposed the following gas equation to allow for non-ideal behaviour:

$$(p + a/V_m^2)(V_m - b) = RT$$

The constant a allows for the intermolecular forces of attraction and the constant b allows for the volume of the molecules themselves.

Exam practice (10 minutes)   answers: pages 24–5

(a) Calculate the volume occupied by 0.500 g of propanone at 100 °C and 101.3 kPa pressure.

(b) The equation for the dissociation of dinitrogen tetraoxide is

$$N_2O_4(g) \rightleftharpoons 2NO_2(g)$$

The dinitrogen tetraoxide is 20% dissociated at a temperature of 28 °C and 101.3 kPa pressure. Calculate the volume occupied by one mole of dinitrogen tetraoxide under these conditions.

[The molar gas volume is 22.4 dm^3 at s.t.p.]

Grade booster

Make sure you learn your organic formulae. You should at least know the name and formula of the first member of the important homologous series. Propanone is the simplest ketone. Its traditional name is acetone.

Structured exam question

answers: page 25

(a) Write the ground state electronic configurations, in terms of s and p electrons, for each of the following isolated species:

    (i)   a chlorine atom ....

    (ii) a calcium ion ....

    (iii) a sulphide ion ....

(b) Draw dot and cross diagrams to represent the bonding in each of the following structural units:

    (i)   ammonia, NH_3

    (ii) carbon dioxide, CO_2

    (iii) sodium oxide, Na_2O

(c) (i)   Describe the difference between the structures of sodium chloride and caesium chloride. Your answer may be either a written description or clearly labelled diagrams.

    (ii) Give a reason why sodium chloride and caesium chloride have different structures.

    ....

    ....

(10 min)

Grade booster

If you are drawing a diagram for *simple* ions do *not show* dots and crosses being shared. You won't get a mark.

Grade booster

Simple clearly labelled neat drawings are better than long illegible muddled written descriptions.

Grade booster

They have because they have is true but not the answer. Think ion sizes.

Answers
Atoms, Ions and molecules

Formulae and equations

Checkpoints

1 SO_2
2 (a) K^+I^- (b) $Ca^{2+} SO_4^{2-}$
   (c) $(Li^+)_2O^{2-}$ (d) $Ba^{2+}(OH^-)_2$
   (e) $(Al^{3+})_2(SO_4^{2-})_3$
3 (a) Ar (b) Cl_2 (c) O_3
4 (a) ICl (b) SCl_2 (c) CO (d) NI_3

Exam practice

(a) $KOH + HNO_3 \rightarrow KNO_3 + H_2O$
   $H^+(aq) + OH^-(aq) \rightarrow H_2O(l)$
(b) $Ba(OH)_2 + 2HCl \rightarrow BaCl_2 + 2H_2O$
   $H^+(aq) + OH^-(aq) \rightarrow H_2O(l)$

Atoms and isotopes

Checkpoints

1 Carbon C
2 $^{13}_{6}C$
3 44 and 46
4 (a) 2p, 2n
   (b) (i) Z decreases by 2 and A decreases by 4.
       (ii) Z increases by 1 and A remains the same.
5 Carbon contains small amounts of C-13 and C-14 isotopes.

Exam practice

(a) (i) Cl-35 17p Cl-37 17p (ii) Cl-35 18n Cl-37 20n
(b)

35	35.5	36	36.5	37
▲	1	▲	3	▲

three Cl-35 : one Cl-37 = 75% Cl-35 and 25% Cl-37
Or $A_r(Cl) = 35.5$; if x = percentage of Cl-35, then
$35.5 = [35x + 37(100 - x)]/100$; so $x = 75\%$.

Amount of substance and the mole

Checkpoints

1 (a) (i) 1.01 g (ii) 14.0 g (iii) 48.0 g
   (b) 63.01 g mol^{-1}
2 (a) 11.2 dm^3
   (b) 3×10^{23}

Exam practice

(a) $CaCO_3 \rightarrow CaO + CO_2$
(b) (i) 560 kg of calcium oxide
    (ii) 2.4×10^5 dm^3
(c) E.g. Scarring of landscape by mining; production of dust by processing; release of CO_2 on roasting to CaO.

Atomic structure

Checkpoints

1 13, aluminium

2 (a) He, Ne and Ar
   (b) s^1 elements marked blue
   (c) p^5 elements marked red
3 (a) Li $1s^22s^1$ and Na $1s^22s^22p^63s^1$
   (b) F $1s^22s^22p^5$ and Br $1s^22s^22p^63s^23p^63d^{10}4s^24p^5$
   (c) He $1s^2$, Ne $1s^22s^22p^6$ and Ar $1s^22s^22p^63s^23p^6$

Exam practice

$1s^22s^22p^63s^23p^5$
$1s^22s^22p^63s^23p^63d^54s^1$
$1s^22s^22p^63s^23d^{10}4s^24p^3$

Structure and bonding in elements

Checkpoints

1 As the atomic number increases, the number of electrons per atom increases and so the van der Waals forces increase. Consequently, more energy is needed to separate the liquid molecules and turn them into gas.
2 (a) 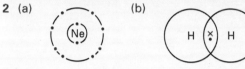 (b)

Exam practice

(a) Malleable; lustrous; good electrical conductivity
(b) Iron becomes harder and less ductile
(c) 8; body-centred cubic

Structure and bonding in covalent compounds

Checkpoints

1 (i)

(ii)

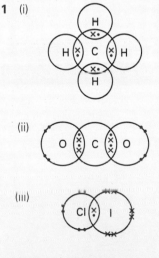

(iii)

2 Fluorine and hydrogen iodide

Exam practice

(a) (i)

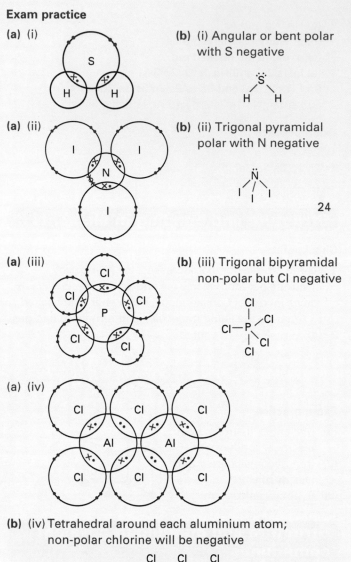

(b) (i) Angular or bent polar with S negative

(a) (ii)

(b) (ii) Trigonal pyramidal polar with N negative

24

(a) (iii)

(b) (iii) Trigonal bipyramidal non-polar but Cl negative

(a) (iv)

(b) (iv) Tetrahedral around each aluminium atom; non-polar chlorine will be negative

---

Structure and bonding in ionic compounds

Checkpoints

1 Cs^+ and F^-

2

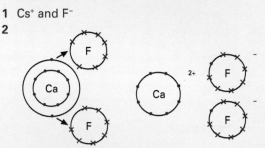

3 (a) Each ion is surrounded by six or eight oppositely charged ions which are held together in the lattice by strong electrostatic forces and much energy is needed to overcome them.

(b) In the solid the ions are in fixed positions but in the liquid the ions are free to move and conduct electricity.

24

---

4 (a) CsCl 0.94 NaCl 0.57

(b) (i) CsCl cubic (double simple interlocking)
    (ii) NaCl cubic (face-centred interlocking)

Exam practice

(a)(i)

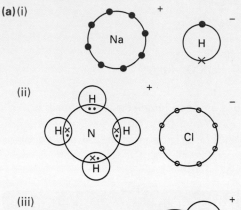

(ii)

(iii)

(b) Aluminium fluoride will have considerable ionic character whereas aluminium chloride is covalent.

(c) The bonding in the carbonate ion will be delocalized (the 2– charge shared equally by the three O-atoms), the three carbon–oxygen bonds will be identical and the C-atom will *not* have a lone pair. Hence the ion will be trigonal planar.

Intermolecular forces

Checkpoints

1 The difference in electronegativity produces a permanent dipole in I–Cl. The Br–Br cannot have a permanent dipole.

2 (a) This is the enthalpy change when one mole of sodium chloride is dissolved in water to a given dilution. In data books the values are usually given to infinite dilution.

(b) Endothermically.

3 (a) Hydrogen sulphide H_2S
    Hydrogen selenide H_2Se
    Hydrogen telluride H_2Te
    Hydrogen chloride HCl
    Hydrogen bromide HBr
    Hydrogen iodide HI
    Phosphine PH_3
    Arsine AsH_3
    Stibine SbH_3
    Silane SiH_4
    Germane GeH_4
    Stannane SnH_4

(b) The average number of hydrogen bonds between a pair of water molecules is two but it is only one between a pair of HF or NH_3 molecule.

Exam practice

(a)(i) Fluorine is the most electronegative element so the hydrogen bonding between HF molecules outweighs the increase in van der Waals forces with increasing molar mass from HCl to HI. Extra energy is needed to overcome the hydrogen bonding before the $H_2F_2(l)$ molecules can separate and become gaseous.

(ii) The dimer is held together by hydrogen bonds:

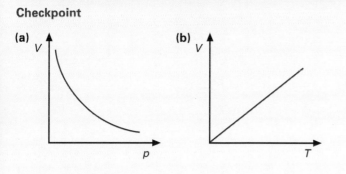

(iii) In ice there is an open lattice structure which is held together by hydrogen bonding. On melting, the hydrogen bonding begins to break and the open structure starts collapsing. The water molecules begin to pack more closely together in the liquid state so the density increases until it reaches a maximum at 4 °C.

(b) (i) Two interlocking simple cubic systems giving a coordination number of 8 : 8.

(ii) The radius ratio for sodium chloride is smaller than that for caesium chloride, so the sodium and chloride ions can form a face-centred cubic lattice with a coordination number of 6 : 6.

Gases

Checkpoint

(a)

(b)

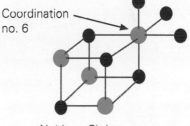

Exam practice

(a) Molar mass of propanone is 58.08 g mol^{-1} so 0.500 g is 0.500/58.08 = 0.008 61 mol propanone. At 273 K the volume of one mole is 22 400 cm^3 so for 0.008 61 mol V is 22 400 × 0.008 61 = 192.9 cm^3 At 373 K the volume is 192.9 × (373/273) = 263.4 cm^3 so volume of propanone will be 263 cm^3.

(b) One mole of ideal gas at 28 °C and 101.3 kPa would be 22.4 × (301/273) = 24.7 dm^3.
When 20% of the $N_2O_4(g)$ dissociates the total amount of gas molecules increases by 20% (because for every $N_2O_4(g)$ molecule lost, two $NO_2(g)$ molecules are gained). So volume increases by 20% from 24.7 dm^3 to 29.6 dm^3.

Structured exam question

(a) (i) a chlorine atom $1s^2 2s^2 2p^6 3s^2 3p^5$
(ii) a calcium ion $1s^2 2s^2 2p^6 3s^2 3p^6 4s^2$
(iii) a sulphide ion $1s^2 2s^2 2p^6 3s^2 3p^6$

(b) (i)   ammonia, NH_3

(ii)   carbon dioxide, CO_2

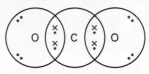

(iii)   sodium oxide, Na_2O

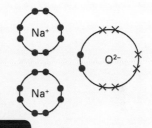

(c) (i) In sodium chloride, each ion is surrounded by six of the oppositely charged ions. Sodium chloride has a double interlocking face-centred cubic structure. In caesium chloride, each ion is surrounded by eight

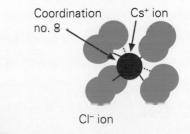

Coordination no. 6

Na$^+$ ion   Cl$^-$ ion

of the oppositely charged ions. Caesium chloride has a double interlocking simple cubic structure.

Coordination no. 8   Cs$^+$ ion

Cl$^-$ ion

(ii) The caesium ion is much larger than the sodium ion and the radius ratio for caesium chloride is larger than for sodium chloride.

Energetics 1

In this chapter you will meet the idea of an enthalpy change ΔH, a heat change at constant pressure. When you see the symbol $\Delta H^\ominus$ you will know this refers to a change under standard conditions of temperature, pressure, etc. For A-level you should know how to measure heats of reactions such as combustion and neutralization. Make sure you understand and can use Hess's law to calculate enthalpy changes of formation, bond enthalpies and lattice energies. These calculations are very popular in exams.

By the end of this chapter you will be able to

- Calculate enthalpy changes from experimental data. Use temperature rises and the equation '$H = mc\Delta T$' to calculate the heat produced in a calorimeter

- Use Hess's law in thermochemical calculations

- Calculate enthalpy changes using ΔH_f (enthalpy of formation), ΔH_c (enthalpy of combustion) and bond energy data

- Draw and use Born–Haber cycles

Topic checklist

	Edexcel		AQA		OCR		WJEC		CCEA	
	AS	A2	AS	A2	AS	A2	AS	A2	AS	A2
Measuring enthalpy changes	O		O		O		O		O	
ΔH and Hess's law	O		O		O		O		O	
Bond energies	O		O		O		O		O	
Lattice energies		●		●		●		●		●

Tick each of the boxes below when you are satisfied that you have mastered the topic

Measuring enthalpy changes

Chemists use a bomb calorimeter to measure heats of combustion at constant volume. You could measure heats of combustion at constant pressure with a flame calorimeter and heat changes involving solutions with a polystyrene cup. In every experiment the basic principles are the same; what are they?

The jargon

Heat capacity means the amount of heat per degree temperature change.
Specific heat capacity means the amount of heat per degree temperature change per unit mass.

Watch out!

If you want the top grade don't confuse enthalpy change ΔH (heat change at constant pressure measured in a flame calorimeter) with internal energy ΔU (heat change at constant volume measured in a bomb calorimeter).

Examiner's secret

You can look up the specific heat capacity of water (4.2 J K⁻¹ g⁻¹) in a data book. You will be given the value in an examination.

Checkpoint 1

Calculate the molar enthalpy change of neutralization for hydrochloric acid and sodium hydroxide from the information given, assuming the temperature rise to have been 6.9°C.

Don't forget!

Always give your final answer to the appropriate number of significant figures.

Grade booster

You will often be asked to explain why an experiment is inaccurate and to suggest ways of making it more accurate. The experiment is usually one you could have done yourself. Be prepared if you want the top grade.

Basic principles

To determine a heat change (δh) involving a certain amount of substance, we measure the temperature change (δT) in the contents of a calorimeter with heat capacity C: $\delta h = \delta T \times C$. We can write this equation in words and with units:

(heat change/kJ) = (temperature change/K) × (heat capacity/kJ K⁻¹)

units: kilojoules   Kelvin   kilojoules per Kelvin

K cancels K⁻¹

We use the measured δh to calculate ΔH, the molar heat change.

→ An enthalpy change ΔH is a heat change **at constant pressure**.
→ For **exothermic** processes ΔH has a *negative* value.
→ For **endothermic** processes ΔH has a *positive* value.

Using a simple calorimeter

Example: Measuring the **enthalpy change of neutralization**

1   React 50 cm³ hydrochloric acid with 50 cm³ aqueous sodium hydroxide, each solution of concentration 1.0 mol dm⁻³.
2   Measure the rise in temperature of the resulting aqueous sodium chloride.
3   Ignore the heat capacity of the cup and assume the specific heat capacity of the aqueous sodium chloride to be that of water.
4   Multiply 4.2 J K⁻¹ g⁻¹ by the mass (g) of the sodium chloride solution to find its heat capacity.
5   Calculate the **molar** enthalpy change of neutralization.

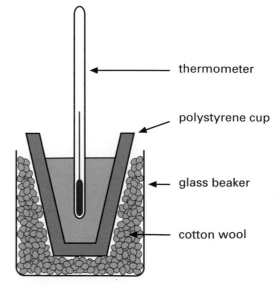

- thermometer
- polystyrene cup
- glass beaker
- cotton wool

Using a flame calorimeter

A flame calorimeter measures heats of combustion *at constant pressure*.

Example: Measuring the enthalpy change of combustion of an alcohol.

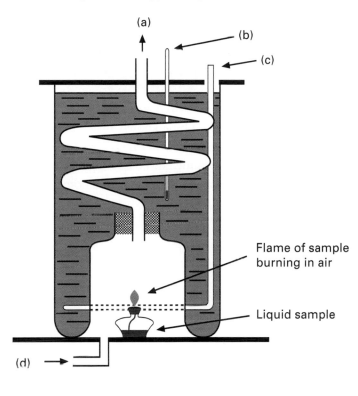

Flame of sample burning in air

Liquid sample

(d)

To determine the heat capacity (C) of the calorimeter you could, for example, measure the loss in mass of a sample of propan-2-ol burnt in the calorimeter and the resulting temperature rise. You could then calculate the value of C using the expression

$$\left\{ \frac{\Delta H}{\text{kJ mol}^{-1}} \times \left[\left(\frac{\delta m}{\text{g}} \right) \div \left(\frac{M}{\text{g mol}^{-1}} \right) \right] \right\} \div \left(\frac{\delta T}{\text{K}} \right)$$

In practice it is quite difficult to obtain a steady, smokeless flame simply by adjusting the height of the burner's wick.

answer: page 37

Exam practice (8 min)

In an experiment to measure the heat of combustion of butan-1-ol ($M_r = 74.1$) a student first determined the heat capacity of a flame calorimeter using propan-2-ol ($M_r = 60.1$) whose enthalpy change of combustion is 2 006 kJ mol^{-1}. In the calibration experiment the combustion of 0.149 g propan-2-ol produced a temperature rise of 5.42 °C. In a second experiment the combustion of 0.152 g butan-1-ol produced a temperature rise of 5.21 °C.

(a) Calculate (i) the heat capacity of the flame calorimeter and (ii) the student's value for the heat of combustion of the butan-1-ol.

(b) The standard enthalpy of combustion of butan-1-ol is 2 676 kJ mol^{-1}. Comment on the accuracy of the student's experiment and suggest two ways of improving it.

ΔH and Hess's law

Molar heat changes vary with temperature, pressure and the concentration of the reactants. In order to compare different reactions, we require heat changes under standard conditions. According to Hess's law, the standard heat change will be the same regardless of how the reaction is carried out. But first, another calorimeter.

The jargon

Bomb refers to the fact that the solid or liquid sample is ignited electrically in an atmosphere of pure oxygen at a pressure of about 20 atm.

Using a bomb calorimeter

We find the heat capacity (C) of a bomb calorimeter by measuring the temperature rise (δT) produced by a known amount of heat (δh): $C = \delta h/\delta T$. This heat could be supplied either by a measured amount of electrical energy or by burning a measured amount of a reference substance such as benzoic acid with a known heat of combustion.

Checkpoint 1

Fill in the four missing diagram labels (a) to (d) to explain the purpose of each part being labelled.

The jargon

Internal energy change ΔU is the heat change at constant volume. We calculate ΔH from ΔU using $\Delta H = \Delta U + \Delta nRT$ where R is the gas constant, T is the absolute temperature and Δn is the number of moles of gaseous products minus the number of moles of gaseous reactants.

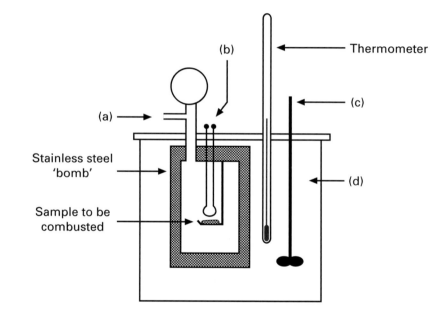

(b)

Thermometer

(a)

(c)

Stainless steel 'bomb'

(d)

Sample to be combusted

A bomb calorimeter measures what we call **internal energy changes** of combustion (ΔU_c). Internal energy change is used to calculate the enthalpy change ΔH_c.

→ *Combustion* is always *exothermic* so ΔH_c is always *negative*.

Checkpoint 2

Calculate the molar enthalpy of combustion of octane, C_8H_{18}, given that 0.114 g of the liquid gives off 5.47 kJ of heat when burnt completely to carbon dioxide and water.

Molar enthalpy change

The amount of substance used in calorimeters is quite small: much less than one mole. The heat change found for a measured but small amount of reaction can be used to calculate the molar enthalpy change.

→ The **molar enthalpy change** of combustion is the heat evolved when one mole of substance is completely burnt in oxygen.

Watch out!

Molar can refer to the 'amount of change' specified by the chemical equation representing the amounts of substances involved in the change.

Standard molar enthalpy change

Molar enthalpy change is a more precise term than molar heat change but it is still not precise enough. The ΔH value for a reaction may vary with temperature, pressure, concentration and physical state of reactants and products.

→ $\Delta H^{\ominus}_{298}$ represents a **standard molar enthalpy change** for a process in which all substances are in their most stable forms at a temperature of 298 K (25 °C) and a pressure of 101 kPa (1 atm) and the concentration of any solution is 1 mol dm^{-3}.

Standard molar enthalpy change of combustion

→ $\Delta H^{\ominus}_{c,298}$ represents the heat change at *constant pressure* when one mole of the substance (at 298 K and 1 atm) is *completely* burnt to form products (at 298 K and 1 atm).

Enthalpies of combustion are particularly important for organic compounds and may be shown with an equation. For example:

$$C_4H_9NH_2(l) + 6\tfrac{3}{4}O_2(g)$$
$$\rightarrow 4CO_2(g) + 5\tfrac{1}{2}H_2O(l) + \tfrac{1}{2}N_2(g); \Delta H^{\ominus}_{c,298} = -3\,018 \text{ kJ mol}^{-1}$$

The 3 018 kJ of heat released to the surroundings includes $5\tfrac{1}{2} \times 41$ kJ of heat released when the water condenses to a liquid because

$$H_2O(g) \rightarrow H_2O(l); \Delta H^{\ominus} = -41 \text{ kJ mol}^{-1}$$

Standard molar enthalpy change of formation

→ $\Delta H^{\ominus}_{f,298}$ represents the heat change at *constant pressure* when one mole of the substance (at 298 K and 1 atm) is formed from its constituent elements in their most stable form at 298 K and 1 atm.

The enthalpy change of combustion of some elements is the same as the enthalpy change of formation of their oxides. For example:

$$C(\text{graphite}) + O_2(g) \rightarrow CO_2(g); \Delta H^{\ominus}_{298} = -393.5 \text{ kJ mol}^{-1}$$

Hess's law

According to the first law of thermodynamics

→ energy cannot be created and cannot be destroyed

German Henri Hess stated his law of constant heat summation in 1840. It is a particular case of the more general first law of thermodynamics. **Hess's law** has been expressed in many ways:

→ the standard molar enthalpy change of a process is independent of the means or route by which that process takes place

Exam practice (4 min)   answer: page 37

Use the following standard enthalpy changes of combustion to determine the standard enthalpy change of formation of methane. Include the sign in your answer and state whether methane is an exothermic or endothermic compound.

Substance

Standard enthalpy change of combustion/kJ mol^{-1}

Hydrogen, $H_2(g)$	−286
Carbon, C(graphite)	−394
Methane, $CH_4(g)$	−890

The jargon

Standard enthalpies of combustion are listed in data books under the column heading: $\dfrac{\Delta H^{\ominus}_{c,298}}{\text{kJ mol}^{-1}}$

Watch out!

The complete combustion of nitrogen-containing organic compounds does not produce oxides of nitrogen.

Don't forget!

Many compounds cannot be formed by direct combination of their elements. In those cases $\Delta H^{\ominus}_{f,298}$ must be found indirectly using Hess's law.

The jargon

Thermodynamics is the branch of science dealing with energy changes, especially heat changes, that occur during physical and chemical processes.

Watch out!

Strictly speaking, Hess's law breaks down if, as well as enthalpy changes, the processes involve other forms of energy changes.

Grade booster

You should know how to use Hess's law to calculate enthalpy changes that cannot be determined by direct experimental measurement. You will get marks for drawing a clearly labelled energy cycle diagram. Bear this in mind if you want the top grade.

Bond energies

When a covalent bond forms between two atoms, energy is given out and the process is exothermic. When a covalent bond between two atoms is broken, energy is taken in and the process is endothermic. The amount of energy involved in these processes depends upon the nature of the atoms and the bonds that join them.

Don't forget!

Bond breaking is endothermic
ΔH is positive
Bond formation is exothermic
ΔH is negative.

The jargon

Strictly speaking, bond energy refers to an internal change (heat change at constant volume) and is not quite the same as bond enthalpy. The difference is small and not important at A-level.

Bond dissociation enthalpies

The difference in the standard molar enthalpies of combustion of the lower alkanes and of the primary alcohols is about $650 \, kJ \, mol^{-1}$. The difference in molecular formula from one alkane (or alcohol) to the next is CH_2. So the $650 \, kJ$ corresponds to the combustion of a CH_2 group to form CO_2 and H_2O.

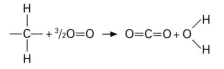

The jargon

Bond dissociation enthalpy applies to a specific bond in a specific molecule.

We suppose the $650 \, kJ$ represents the difference between the energy taken in by the breaking of the C—H, O=O bonds and the energy given out by the forming of the C=O and O—H bonds. It follows that bonds need definite energies to break them. Here are the first and second standard bond dissociation enthalpies of water:

first $H{-}OH(g) \rightarrow H(g) + OH(g)$; $\Delta H^{\ominus} = +498 \, kJ \, mol^{-1}$
second $O{-}H(g) \rightarrow O(g) + H(g)$; $\Delta H^{\ominus} = +430 \, kJ \, mol^{-1}$

→ The precise strength of a bond (X—Y) depends upon the other atoms or groups of atoms attached to X and Y.

Checkpoint 1

The H—O bond energy, E(H—O), listed in a data book is the average value of the first and second standard bond dissociation enthalpies of water. Calculate E(H—O) in $kJ \, mol^{-1}$.

Bond energy

In a data book the H—O bond energy is listed as $E(H{-}O) = +464 \, kJ \, mol^{-1}$ which is the average value of the first and second standard bond dissociation enthalpies of water. $E(C{-}H) = 435 \, kJ \, mol^{-1}$ is an average value of the standard bond dissociation enthalpies of the four bonds in methane: namely $CH_3{-}H$, $CH_2{-}H$, $CH{-}H$, $C{-}H$.

Checkpoint 2

Write balanced equations for the dissociation of each of the four bonds in methane.

→ Bond energy is the *average* standard enthalpy change for the breaking of a mole of bonds in a gaseous molecule to form gaseous atoms.
→ Bond energies indicate the strength of the forces holding together atoms in a covalent molecule.

Calculating bond energies with Hess's law

You should understand how to use standard molar enthalpy changes of formation of molecules and of atomization of elements to calculate an average standard molar bond enthalpy.

Grade booster

You are expected to know that most bond energy values are average (or mean) values but you are not expected to know the experimental techniques for measuring them. Bear this in mind if you want the top grade.

→ The standard molar enthalpy change of **atomization** of an element is the enthalpy change when *one mole of gaseous atoms* are formed from the element under standard conditions.

Example:
Calculating
$E(C-H)$ for
methane

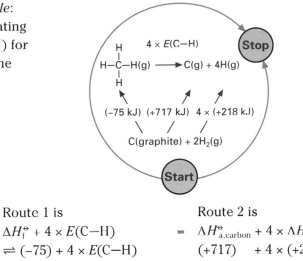

Route 1 is

$\Delta H_f^\ominus + 4 \times E(C-H)$
$\rightleftharpoons (-75) + 4 \times E(C-H)$
$\rightleftharpoons E(C-H) = +416 \text{ kJ mol}^{-1}$

Route 2 is

$= \Delta H_{a,\text{carbon}}^\ominus + 4 \times \Delta H_{a,\text{hydrogen}}^\ominus$
$(+717) + 4 \times (+218)$

Watch out!

For atomization of elements, molar refers to one mole of atoms produced: $^1/_2 H_2(g) \rightarrow H(g)$, $\Delta H = +218 \text{ kJ mol}^{-1}$ It does *not* refer to one mole of element being atomized!
$H_2(g) \rightarrow 2H(g)$; $\Delta H = +436 \text{ kJ mol}^{-1}$

Patterns in bond energies

→ Bond energies range from 150 kJ mol^{-1} for weak bonds ($HO-OH$, $I-I$, $F-F$, H_2N-NH_2) to $350-550 \text{ kJ mol}^{-1}$ for strong bonds ($C-C$, $C-H$, $N-H$, $O-H$, $H-Cl$) to around $1\,000 \text{ kJ mol}^{-1}$ for very strong bonds ($N\equiv N$, $C\equiv C$, $C\equiv O$).

→ Bond energy increases with the number of shared electron-pairs ($E(C-C) = +347$, $E(C=C) = +612$, $E(C\equiv C) = +838 \text{ kJ mol}^{-1}$).

Calculations using bond energies

You could be asked to calculate an approximate value for the enthalpy change of a reaction.

Example: Estimate the enthalpy change for the catalytic reforming of hexane to cyclohexene and hydrogen which is part of the process for the industrial production of petrol: $C_6H_{14} \rightarrow C_6H_{12} + H_2$.

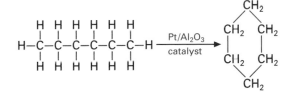

Break 2 C—H bonds
$\rightleftharpoons 2 \times 413 \text{ kJ used up}$
$\rightleftharpoons 826 \text{ kJ used up}$
$\rightleftharpoons$ net release of $1\,130 - 826 = 304 \text{ kJ}$
$\rightleftharpoons \Delta H_f^\ominus = -304 \text{ kJ mol}^{-1}$

Form 2 C—C bonds and
1 H—H bond
2×347 and $1 \times 436 \text{ kJ given out}$
$1\,130 \text{ kJ given out}$

Checkpoint 3

(a) Use the following data to calculate the approximate enthalpy of formation of the hydrogen halides.
$H_2(g) \rightarrow 2H(g)$; $\Delta H = +436 \text{ kJ mol}^{-1}$

Element X	$E(X-X)$ $/ \text{kJ mol}^{-1}$	$E(H-X)$ $/ \text{kJ mol}^{-1}$
Fluorine	158	562
Chlorine	244	431
Bromine	194	366
Iodine	152	299

(b) Comment of the reactivity of the halogens in relation to their $E(X-X)$ values and the enthalpy of formation of the hydrogen halides.

Exam practice (10 min)   answer: page 37

(a) Calculate the mean bond enthalpy $E(N-H)$ for ammonia. $\Delta H_f^\ominus$ for ammonia = $-46.1 \text{ kJ mol}^{-1}$ and the enthalpy changes of atomization of nitrogen and hydrogen are 473 and 218 kJ mol^{-1} respectively.

(b) How do the bond energy, bond length and thermal stability of the hydrogen halides vary with increasing molar mass from HF to HI?

Grade booster

If you want the top grade you should check your calculations carefully and show enough working for the examiner to follow your reasoning.

Lattice energies

The jargon

The term *lattice* is used to describe a regular array of particles producing a solid, e.g. ions form ionic lattices, molecules form molecular lattices.

The jargon

Strictly speaking, electron affinity is the molar internal energy change for gaseous atoms (or ions) gaining an electron and is not quite the same as electron gain enthalpy $\Delta H_e^{\ominus}$. The distinction is not significant at A-level.

Watch out!

If you want the top grade do NOT confuse the imaginary lattice enthalpy change with the real enthalpy of formation of the compound from its elements in the normal states!

Don't forget!

The *first* electron affinity (gain enthalpy) is usually negative and the process is exothermic.

Checkpoint 1

(a) Suggest why the *second* electron affinity of, for example, oxygen is positive and the process is endothermic.
(b) State, and suggest an explanation for, the trend in the electron affinity with atomic mass of Cl, Br and I.

Grade booster

When you do these types of calculation, keep the sign with its value inside a bracket so you don't lose track of it. Put the sign with your final answer even if the result is positive. You get a mark for the sign! Bear this in mind if you want the top grade.

When a metal and non-metal react to form an ionic compound, the process is exothermic. We see this energy output as the balance between the energy used up to turn the elements into gaseous ions and the energy given out when the solid lattice forms from the gaseous ions.

Energy cycles

Solid sodium (body-centred metallic crystal) and gaseous chlorine (simple diatomic molecules) react to form sodium chloride (face-centred ionic crystal): $Na(s) + \frac{1}{2}Cl_2(g) \rightarrow NaCl(s)$; $\Delta H_f^{\ominus} = 411 \text{ kJ mol}^{-1}$.
We can imagine the process involving the following steps.

1 Sodium vaporizes and the gaseous atoms lose electrons:

$$Na(s) \rightarrow Na(g); \Delta H_{at}^{\ominus} = +107 \text{ kJ mol}^{-1}$$
$$Na(g) \rightarrow e^- + Na^+(g); \Delta H_i^{\ominus} = +496 \text{ kJ mol}^{-1}$$

2 Chlorine molecules dissociate and the gaseous atoms gain electrons:

$$\frac{1}{2}Cl_2(g) \rightarrow Cl(g); \Delta H_i^{\ominus} = +122 \text{ kJ mol}^{-1}$$
$$Cl(g) + e^- \rightarrow Cl^-(g); \Delta H_e^{\ominus} = -349 \text{ kJ mol}^{-1}$$

3 Gaseous sodium cations and chloride anions form a crystal lattice:

$$Na^+(g) + Cl^-(g) \rightarrow NaCl(s); \Delta H_l^{\ominus}$$

→ Lattice energy is the exothermic heat of formation of one mole of an ionic solid from its constituent *gaseous ions*.

We can draw an energy cycle for these steps and use Hess's law to calculate a value for the lattice energy. This is usually called a Born–Haber cycle.

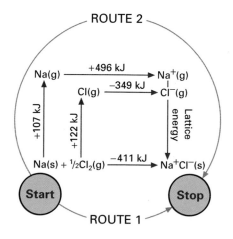

Route 1 = Route 2
$(-411) = (+107) + (+496) + (+122) + (-349) + \Delta H_e^{\ominus}$
$\Rightarrow \Delta H_e^{\ominus} = (-411) - (+107) - (+496) - (+122) - (-349)$
$\Rightarrow \qquad = -787 \text{ kJ mol}^{-1}$

Born–Haber cycles

It is instructive to present energy cycles like the **Born–Haber cycle** to show the energy changes against a vertical scale.

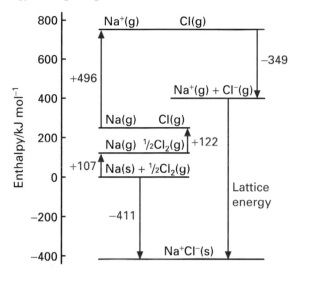

Lattice energy values

You will not have to remember specific lattice energy values but you should have some idea of their range and magnitude compared to bond energies.

→ Lattice enthalpies range from around 600 kJ mol⁻¹ to more than 4 000 kJ mol⁻¹ which is well beyond the strongest covalent bond.

You should remember and be able to suggest explanations for certain patterns in the values of lattice energies.

→ Lattice energies for doubly charged ions are usually much more exothermic than those for singly charged ions.
→ Lattice energies become less exothermic as the size of the anion increases.
→ Lattice energies become less exothermic as the size of the cation increases.

You should appreciate how experimental and theoretical lattice energies indicate the character of alkali metal and silver halides.

→ Alkali metal halides are ionic with little or no covalent character.
→ Silver halides have a considerable percentage of covalent character.

Exam practice (10 min)   answer: page 38

Ethene, C_2H_4, burns completely in oxygen to form carbon dioxide and water.
Ethene also reacts with chlorine to form 1,2-dichloroethane, CH_2ClCH_2Cl.
Calculate the enthalpy change of formation

(a) of ethene from its elements, and
(b) of 1,2-dichloroethane from ethene and chlorine.

Use the following molar enthalpies of combustion: $H_2(g)$ –286; $C(s)$ –394;
$C_2H_4(g)$ –1 411, $CH_2ClCH_2Cl(l)$ –1 246 kJ mol⁻¹.

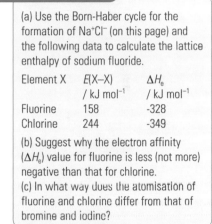

The jargon

The vertical scale represents the enthalpy H (or heat content). For elements in their most stable states, $H = 0$ by definition. Strictly speaking, the diagram does not need the arrow heads and the + or – signs. ↑ and + mean endothermic. ↓ and – mean exothermic.

Checkpoint 2

(a) Use the Born-Haber cycle for the formation of Na^+Cl^- (on this page) and the following data to calculate the lattice enthalpy of sodium fluoride.

Element X	$E(X–X)$ / kJ mol⁻¹	ΔH_e / kJ mol⁻¹
Fluorine	158	-328
Chlorine	244	-349

(b) Suggest why the electron affinity (ΔH_e) value for fluorine is less (not more) negative than that for chlorine.
(c) In what way does the atomisation of fluorine and chlorine differ from that of bromine and iodine?

Watch out!

Some Data Books define lattice enthalpy as dissociation lattice enthalpy and ΔH_{lat} is positive.

Action point

Verify these patterns by looking up the lattice enthalpies for the following: LiF, LiCl, LiBr, LiI, NaF, NaCl, KF, RbF, CsF, Na_2O and $MgCl_2$.

The jargon

An *experimental lattice energy* is determined using a Born–Haber cycle like the one above for NaCl. A *theoretical lattice energy* is a mathematical prediction based on a model of perfectly spherical ions carrying complete charges.

Grade booster

If you want the top grade remember that marks are given for a clearly labelled energy cycle.

Structured exam question

answer: page 38

(a) The standard molar enthalpy change of combustion ($\Delta H^{\ominus}_{c,298}$) of methane is -890.3 kJ mol^{-1}. A value for the enthalpy change of combustion of methane may be calculated from standard molar enthalpy changes of formation ($\Delta H^{\ominus}_{f,298}$) or from bond dissociation energies (E).

(i) Calculate the enthalpy change of combustion of methane using the following values for the standard molar enthalpy changes of formation.

Compound	$\Delta H^{\ominus}_{f,298}$/kJ mol^{-1}
Carbon dioxide	-394
Water	-286
Methane	-74.8

....
....
....

(ii) Calculate the enthalpy change of combustion of methane using the following values for the average bond dissociation energies.

Bond	E/kJ mol^{-1}
C—H	$+435$
O=O	$+498$
C=O	$+805$
H—O	$+464$

....
....
....

(iii) Comment upon the values calculated in (i) and (ii) above.

....
....
....

(b) Calculate the enthalpy change of solution of solid sodium chloride from the following data.

Change	Enthalpy change/kJ mol^{-1}
$NaCl(s) \rightarrow Na^+(g) + Cl^-(g)$	$+787$
$Cl^-(g) + aq \rightarrow Cl^-(aq)$	-377
$Na^+(g) + aq \rightarrow Na^+(aq)$	-406

....
....
....
....
....
....
....
....
....
....

Grade booster

You will impress if you keep the − and + signs with the values in brackets until your final step. Lose signs, you lose marks and the top grade.

Watch out!

Don't forget that breaking bonds uses energy and therefore is endothermic with ΔH being positive.

Examiner's secrets

You may be given lines for writing or space for diagrams and/or calculations. The number of lines and the amount of space is a clue to how much is expected for a good answer *but* the number of marks (1 mark = 1 exam minute) is a better guide.

(10 min)

Answers
Energetics

Measuring enthalpy changes

Checkpoints

1 Heat produced is $100 \times 4.2 \times 6.9 = 2\,898$ J. Amount of acid was $1.0 \times 50/1\,000 = 0.05$ mol HCl(aq) and this reacted with the same amount of NaOH(aq). So the heat produced by 1 mol HCl(aq) or NaOH(aq) would be $2\,898/0.05 = 58\,000$ Js. Hence the molar enthalpy change is -58 kJ mol^{-1}

> **Examiner's secrets**
>
> You could lose marks if you give more than two significant figures in your answer!

2 (a) To pump
   (b) Thermometer
   (c) Stirrer to ensure a uniform temperature
   (d) Air (or oxygen)
3 J K^{-1} or kJ K^{-1}
4 Production of carbon dioxide adding to the greenhouse effect and global warming. Air pollution by carbon monoxide produced by incomplete combustion.

Exam practice

(a)(i) Amount (in moles) of propan-2-ol used is $0.149/60.1 = 2.48 \times 10^{-3}$ mol. Heat produced is $2.48 \times 10^{-3} \times 2\,006$ kJ ($= 4.975$ kJ). This causes a temperature rise of $5.42\,°C$ and so the heat to cause a rise of $1\,°C$ is $(2.48 \times 10^{-3} \times 2\,006)/5.42 = 0.918$ kJ K^{-1}, the heat capacity of the calorimeter.

   (ii) Heat produced by 0.152 g of butan-1-ol is 0.918 kJ K$^{-1} \times 5.21$ K $= 4.78$ kJ.
   Heat produced by one mole ($= 74.1$ g) of butan-1-ol is $4.78 \times 74.1/0.152 = 2\,330$ kJ mol^{-1}.

(b) The student's value is too low. $2\,676 - 2\,330 = 346$ kJ of heat may have been 'lost' to the surroundings or not produced because combustion was incomplete. Surround the apparatus with thermal insulation. Enrich the air with oxygen to ensure more complete combustion. Use a more accurate thermometer. (Marks would be given for any sensible suggestion.)

ΔH and Hess's law

Checkpoints

1 (a) Oxygen at high pressure
   (b) Electrical connection to ignite sample
   (c) Stirrer to ensure a uniform temperature
   (d) Water to take up the heat produced
2 M_r(octane) = 114
   0.114 g is 0.01 mol
   Heat produced = 5.47 kJ
   $\Delta H_c = -5.47 \div 0.001 = -5.470$ kJ mol^{-1}

Exam practice

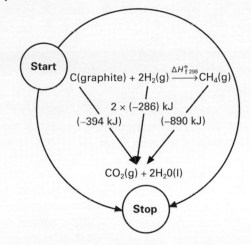

$\Delta H + (-890) = (-394) + (2 \times (-286))$
$\Delta H = +890 - 394 - 572$
$\qquad = -76$ kJ mol^{-1}, an exothermic compound

Bond energies

Checkpoints

1 E(H–OH) $= +498$ kJ mol^{-1} and E(H–O) $= +430$ kJ mol^{-1}
   Total energy for both bonds is $498 + 430 = 928$ kJ mol^{-1}
   So average bond energy is $928 / 2 = +464$ kJ mol^{-1}
2 first H–CH$_3$(g) → H(g) + CH$_3$(g)
   second   H–CH$_2$(g) → H(g) + CH$_2$(g)
   third   H–CH(g) → H(g) + CH(g)
   fourth   H–C(g) → H(g) + C(g)
3(a) $\frac{1}{2}$H–H(g) + $\frac{1}{2}$X–X(g) → H–X(g)
   ΔH_f is $\frac{1}{2}E$(H–H) $+ \frac{1}{2}E$(X–X) $- E$(H–X) in kJ mol^{-1}

Fluorine	218	+	79	-	562	=	-265
Chlorine	218	+	122	-	431	=	-91
Bromine	218	+	97	-	366	=	-51
Iodine	218	+	76	-	299	=	-5

3(b) The ΔH_f values are in accord with the decreasing reacivity with increasing atomic number. The very large (negative) value of ΔH for HF stems from the comparatively weak F–F bond and strong H–F bond

Exam practice

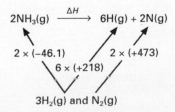

(a) Atomization enthalpy of hydrogen is $+218$ kJ mol^{-1}. From the cycle
   $$2 \times (-46.1) + \Delta H = 6 \times (+218) + 2 \times (+473)$$
   so $-92.2 + \Delta H = 1\,308 + 946$

   Hence $\Delta H = 2\,346.2$ kJ. This is for breaking six moles of N–H bonds. Therefore E(N–H) is $2\,346/6 = 391$ kJ mol^{-1}.

(b) From HF to HI the bond energy decreases and therefore the thermal stability decreases. Bond length increases from HF to HI.

Lattice energies

Checkpoint 1

(a) Adding an electron to the O^- ion uses energy to overcome the repulsion between the negative charges so the enthalpy change for $O^-(g) + e^- \rightarrow O^{2-}(g)$ must be positive.

(b) The electron affinities become less exothermic from Cl to I. The nuclear charge increases from Cl to I but its attraction for the added electron does not increase because the inner-shell shielding also increases from Cl to I. The increase in shielding outweighs the increase in nuclear charge so the attraction for the added electron decreases from Cl to I and less energy is released.

Checkpoint 2

(a) difference in energy input is $(244/2) - (158/2)$
= 43 kJmol^{-1} less
difference in energy output is $(-349) - (-328)$
= -21 kJmol^{-1} less
net difference in lattice energy
= -22 kJmol^{-1} (i.e. more released)
hence lattice enthalpy NaF is $(-787) + (-22)$
= -807 kJmol^{-1}

(b) Although the electron added to fluorine will be closer to the nucleus than that added to chlorine, the fluorine atom is much smaller than a chlorine atom and the repulsion of the electrons already present will reduce the attraction of the nuclear charge and therefore the electron affinity.

(c) In their standard states. F_2 and Cl_2 are already gaseous but bromine is a liquid and iodine a solid. So the atomisation of Br_2 and I_2 also involves their enthalpies of vaporisation.

Exam practice

(a)

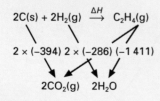

$\Delta H + (-1\,411) = 2 \times (-394) + 2 \times (-286)$
$\Delta H = -(-1\,411) + 2 \times (-394) + 2 \times (-286)$
$\Delta H = +51.0$ kJ mol^{-1}

(b)

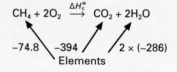

$\Delta H + (-1\,264) = -1\,411$
$\Delta H = -165$ kJ mol^{-1}

Structured exam question

(a) **(i)** Using an energy cycle:

$$CH_4 + 2O_2 \xrightarrow{\Delta H_c^\ominus} CO_2 + 2H_2O$$

−74.8 −394 $2 \times (-286)$
Elements

$-74.8 + \Delta H_c^\ominus = (-394) + 2 \times (-286)$
Hence $\Delta H_c^\ominus = -891$ kJ mol^{-1} 3 sig fig

(ii) $CH_4 + 2O_2 \rightarrow CO_2 + 2H_2O$
$4 \times (+435) + 2 \times (+498)$ $2 \times (+805) + 2 \times [2 \times (+464)]$
So $\Delta H_c^\ominus$ is
$-\{2 \times (+805) + 2 \times [2 \times (+464)] - [4 \times (+435)$
$+ 2 \times (+498)]\} = -\{[1\,610 + 1\,856] - [1\,740 + 996]\}$
= −730 kJ mol^{-1}

(iii) The value −891.2 kJ mol^{-1} should be very close to the experimental value because the standard molar enthalpies of formation are obtained indirectly from experimentally determined enthalpies of combustion. The value −730 kJ mol^{-1} could be expected to differ from −891.2 kJ mol^{-1} because the bond dissociation energies are average energy values and may be determined by interpretation of data such as those obtained from spectroscopy.

(b)

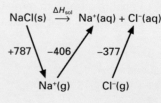

$\Delta H_{sol} = (+787) + (-406) + (-377)$
So $\Delta H_{sol} = + 4$ kJ mol

Kinetics

You probably know how to make hydrochloric acid react faster with magnesium: use more concentrated acid, use magnesium powder instead of ribbon and warm the mixture. In this chapter you will learn more about rates of reactions and how they depend upon concentration and temperature. You will see how collision theory explains this dependence. You will also discover how catalysts work. Make sure you understand the maths behind an expression like 'rate = $k[A]^m[B]^n$'; it's easier than it looks.

By the end of this chapter you will be able to

→ Describe a way of following the rate of a specified reaction

→ Describe how and why temperature, concentration and catalyst affect the rate of a reaction and use the concept of activation energy and the Maxwell–Boltzmann distribution to explain the effects

→ Describe the way in which homogeneous and heterogeneous catalysts work and give examples of homogeneous and heterogeneous catalysts in inorganic and organic chemistry

→ Describe what is meant by a rate equation and, using supplied data, work out the order of reaction, the rate equation, the rate constant and its units

→ Work out a possible mechanism from a rate equation or select which of several supplied mechanisms best fits the rate equation

→ Draw and interpret energy level diagrams (reaction profiles) to describe uncatalysed and catalysed reactions

Topic checklist

Tick each of the boxes below when you are satisfied that you have mastered the topic

	Edexcel		AQA		OCR		WJEC		CCEA	
	AS	A2	AS	A2	AS	A2	AS	A2	AS	A2
Rates of reaction	○	●	○	●	○	●	○	●	○	●
Orders of reaction and mechanisms		●		●		●		●		●
Reaction rates and temperature		●		●		●		●		●
Catalysis		●		●		●		●		●

Rates of reaction

During any chemical reaction the concentrations of reactants decrease and the concentrations of products increase. Rates of reaction are measures of how these concentrations change with time.

Rate of reaction

The rate of a reaction is the change of concentration of a reactant (or of a product) with time.

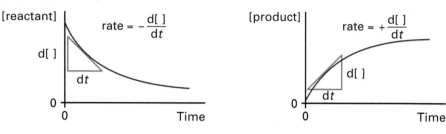

$$\text{rate} = -\frac{d[\ ]}{dt}$$

$$\text{rate} = +\frac{d[\ ]}{dt}$$

Units of rate are usually (but not always) $mol\,dm^{-3}\,s^{-1}$.

Measuring rates of reaction

We follow reactions by measuring the concentration of a reactant (or product) or by monitoring changes in a property of the system. The following are important reactions for A-level chemistry. They proceed at a rate that you could conveniently measure in your laboratory.

Example of reaction	Property measured
$Mg(s) + 2HCl(aq) \rightarrow MgCl_2(aq) + H_2(g)$	Gas volume
$CaCO_3(s) + 2HCl(aq) \rightarrow CaCl_2(aq) + H_2O(l) + CO_2(g)$	Mass
$CH_3CO_2C_2H_5(l) + H_2O(l) \rightarrow CH_3CO_2H(aq) + C_2H_5OH(aq)$	Concentration
$CH_3COCH_3(aq) + I_2(aq) \rightarrow CH_3COCH_2I(aq) + HI(aq)$	Colour intensity

Experiment 1

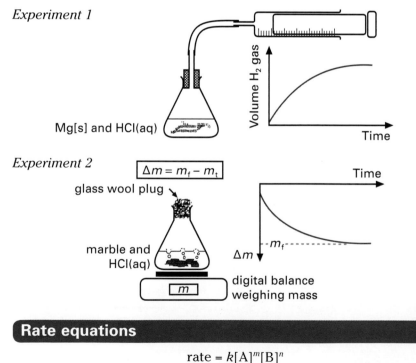

Mg[s] and HCl(aq)

Experiment 2

$\Delta m = m_f - m_t$

glass wool plug

marble and HCl(aq)

Δm ┄ m_f

digital balance weighing mass

Rate equations

$$\text{rate} = k[A]^m[B]^n$$

→ k is the **rate constant** whose value depends upon temperature.
→ m and n are the **orders of reaction** with respect to the reactants.
→ $m + n$ is called the overall order of the reaction.

The jargon

[A] = concentration of A where A is a reactant or a product. d[ ]/dt is the slope (gradient or tangent) at a point on a graph of concentration [ ] against time t and is related to the rate of reaction. Rate = d[A]/dt when A is a product but rate = –d[A]/dt when A is a reactant. The negative sign in front makes the rate values positive.

Grade booster

For the top grade you should know the details of simple experiments you yourself could do to measure reaction rates with standard laboratory apparatus.

Checkpoint 1

(a) State and explain three precautions needed in experiment 1.
(b) Suggest how you could start the reaction and the clock *both together*.

Watch out!

If you want the top grade make sure you understand why the direction of the *y*-axis in experiment 1 is the opposite of the *y*-axis in experiment 2.

Checkpoint 2

(a) State and explain the purpose of the glass wool plug in experiment 2.
(b) If m_t is the reading on the digital balance at time t and m_f is the final reading when the reaction has stopped, what is Δm proportional to?

For A-level chemistry the orders m and n may be 0, 1 or 2 but the overall order $m + n$ will not exceed 2. The associated integrated rate equations express concentration as a function of time.

Watch out!

The value of k *does not* depend upon concentration. k depends only upon the temperature and the reaction chosen.

Zero order reactions

$$-d[\ ]/dt = k_0[\text{reactant}]^0 \quad \text{and} \quad [\text{reactant}] = -k_0t + [\text{reactant}]_{\text{initial}}$$

→ Rate is independent of concentration.
→ The value of k_0 may be obtained from the slope of a graph of concentration against time.
→ The units of k_0 are $\text{mol dm}^{-3}\text{s}^{-1}$.
→ Half-life $= [\ ]/2k_0$.

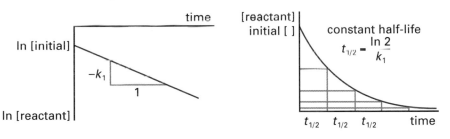

The jargon

Half-life is the time interval for the concentration of a reactant to decrease to half its original value.

The jargon

In means logarithm to the base e. log means logarithm to the base 10.

First order reactions

$$-d[\ ]/dt = k_1[\text{reactant}]^1 \quad \text{and} \quad \ln[\text{reactant}] = -k_1t + \ln[\text{reactant}]_{\text{initial}}$$

→ Half-life $= (\ln 2)/k_1$ is independent of concentration.
→ The value for k_1 may be obtained from the slope of a graph of natural logarithm of concentration against time or from the half-life.
→ The units of k_1 are s^{-1}.

Don't forget!

The decay and half-life of radioactive isotopes provide the best examples of first order reactions.

Don't forget!

The logarithmic equation for first order reactions is the integrated rate equation. This level of mathematical knowledge may not be required by your Awarding Authority. Please check and look at specimen questions.

Second order reactions

$$-d[\ ]/dt = k_2[\text{reactant}]^2 \quad \text{and} \quad 1/[\text{reactant}] = k_1t + 1/[\text{reactant}]_{\text{initial}}$$

→ The value for k_2 may be obtained from the slope of a graph of 1/concentration against time.
→ The units of k_2 are $\text{mol}^{-1}\text{dm}^3\text{s}^{-1}$.
→ Half-life $= 1/(k_2[\ ])$.

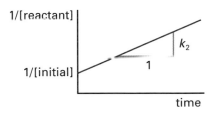

Don't forget!

The *amount* of product formed in any reaction always depends upon the *amount* of reactants even though the rate of reaction may be independent of the concentration of a particular reactant. Reactants always react to produce products!

Exam practice (5 min)   answer: page 49

Carbon-14 has a half-life of 5 730 years and decays to nitrogen-14.

(a) Write an equation for the decay of carbon-14.
(b) How long would it take for the radioactivity of a carbon-14 sample to fall to 1/8 of its original value?
(c) Why is carbon-14 important to archaeologists?

Orders of reaction and mechanisms

The jargon

A *stoichiometric equation* is a balanced chemical equation showing the relative amounts of reactants and products. The (usually simple whole) numbers in front of each formula are known as stoichiometric coefficients.

Checkpoint 1

State and explain two different ways of following the reaction of propanone with iodine.

Watch out!

Drawing tangents to a [ ] vs *t* curve, especially at the start where *t* = 0, can be very inaccurate.

Examiner's secrets

There are often questions asking you to find the orders of a reaction from a table of experimental data.

Don't forget!

Kinetics is a very important branch of physical chemistry. How *fast* reactions proceed (kinetics) is just as important as *how far* reactions proceed (equilibria).

All rate equations must be determined experimentally; they cannot be predicted from the stoichiometric equation for a reaction. Consequently we must do experiments to find the orders and the value of the rate constant for any reaction. The rate equation may reveal the mechanism of the reaction.

Finding the orders of a reaction

The experimentally determined stoichiometry of the acid-catalyzed reaction of propanone and iodine is given by the following equation:

$$CH_3COCH_3(aq) + I_2(aq) \rightarrow CH_3COCH_2I(aq) + H^+(aq) + I^-(aq)$$

To find the orders of reaction with respect to propanone, iodine and acid catalyst we must do several experiments to see how the concentration of each substance alters the rate.

Using initial rates

If you measure the *initial* rate at the start of the reaction (at the moment of mixing the reactants when *t* = 0) then you will also know the *initial* concentrations of all the substances. So if you do several separate experiments doubling the concentration of each substance in turn, you should discover how the concentration of each affects the rate. Here are some typical results:

	Initial concentrations/mol dm^{-3}			*Initial rate*/10^{-5} mol dm^{-3} s^{-1}
Expt	[iodine]	[propanone]	[HCl(aq)]	$-d[I_2(aq)]/dt$
A	0.001	0.5	1.25	1.1
B	0.002	0.5	1.25	1.1
C	0.002	1.0	1.25	2.2
D	0.002	1.0	2.50	4.4

Experiments A and B show that the iodine concentration does *not* affect the rate, so the reaction is *zero order* with respect to [I$_2$(aq)]. Experiments B and C show that the rate doubles when the propanone concentration is doubled, so the reaction is *first order* with respect to [CH$_3$COCH$_3$(aq)]. Experiments C and D show that the rate doubles when the acid concentration is doubled, so the reaction is *first order* with respect to [HCl(aq)]. Consequently the rate equation is

$$-d[I_2(aq)]/dt = k[I_2(aq)]^0[CH_3COCH_3(aq)]^1[HCl(aq)]^1$$

or simply

$$rate = k[CH_3COCH_3(aq)][HCl(aq)]$$

Using half-lives

The variation in half-life with concentration (or time) may indicate the order of a reaction, e.g. for the reaction $N_2O_5(g) \rightarrow N_2O_4(g) + \frac{1}{2}O_2(g)$

Checkpoint 2

For the decomposition of N$_2$O$_5$(g)
(a) write a rate equation
(b) state the order of the reaction and
(c) plot a graph of ln [N$_2$O$_5$] vs time to find a value for *k*

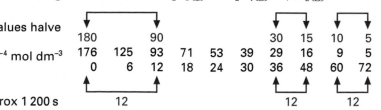

Approx [ ] values halve										
	180		90				30	15	10	5
[N$_2$O$_5$(g)]/10^{-4} mol dm^{-3}	176	125	93	71	53	39	29	16	9	5
Time/10^2 s	0	6	12	18	24	30	36	48	60	72

Half-life approx 1 200 s   12   12   12

Reaction mechanisms

A reaction mechanism is a set of theoretical steps proposed to account for the conversion of reactants into products. For A-level chemistry, a reaction mechanism is usually considered to be a sequence of simple steps, the slowest of which controls the overall rate of the reaction.

→ The slowest step is called **the rate-determining step**.

Iodination of propanone

Here is one plausible mechanism for the propanone–iodine reaction.

$$(CH_3)_2CO(aq) + H_3O^+(aq) \xrightarrow{\text{slow}} (CH_3)_2COH^+(aq) + H_2O(l)$$

$$(CH_3)_2COH^+(aq) + H_2O(l) \xrightarrow{\text{fast}} CH_3-\underset{\underset{OH}{|}}{C}=CH_2(aq) + H_3O^+(aq)$$

$$CH_3-\underset{\underset{OH}{|}}{C}=CH_2(aq) + I_2(aq) \xrightarrow{\text{fast}} CH_3-\underset{\underset{O}{\|}}{C}=CH_2I(aq) + HI(aq)$$

Each step is bimolecular. The (slow) rate-determining step involves one propanone molecule and one hydrogen ion to be consistent with the orders in the rate equation. The second step regenerates the H_3O^+ ion used up in the first step to be consistent with the hydrogen ion being a catalyst. When you combine the equations for the three steps (by adding the left-hand sides together, the right-hand sides together and cancelling identical species on both sides) you get the correct stoichiometric equation for the overall chemical reaction.

Hydrolysis of bromoalkanes

$$C_4H_9Br(l) + H_2O(l) \rightarrow C_4H_9OH(l) + HBr(aq)$$

You could do simple test-tube reactions to show that the hydrolysis of 2-bromo-2-methylpropane is much faster than that of 1-bromobutane.

Bromoalkane RBr	Rate equation	Mechanism
$CH_3-\underset{\underset{CH_3}{\mid}}{\overset{\overset{CH_3}{\mid}}{C}}-Br$	rate $\propto$ [RBr]	S_N1 (a two-step mechanism) $RBr \xrightarrow{\text{slow}} R^+$ $R^+ + OH^- \xrightarrow{\text{fast}} ROH$
$CH_3CH_2CH_2-\underset{\underset{H}{\mid}}{\overset{\overset{H}{\mid}}{C}}-Br$	rate $\propto$ [RBr][OH$^-$]	S_N2 (a one-step mechanism) $RBr + OH^- \xrightarrow{\text{slow}} ROH + Br^-$

Links

See page
148: hydrohalogenation of alkenes;
150: halogenation of alkanes;
152: halogenation of alkenes;
153: nitration of arenes.

The jargon

Bimolecular refers to a step involving *two* reacting species.
Unimolecular refers to a step involving *one* reacting species only.

Action point

Learn the definition of a catalyst and make a list of some important industrial catalysts and the reactions they catalyze.

Checkpoint 3

Combine the equations for the three steps and write down the resulting stoichiometric equation.

Grade booster

Your knowledge of nucleophilic substitution reactions will be tested in organic chemistry but you must be aware of the kinetic evidence for the proposed mechanisms if you want the top grade.

Watch out!

Don't confuse the meanings of S_N1 and S_N2. 1 stands for unimolecular. 2 stands for bimolecular. The 1 and 2 do *not* refer to the number of steps in the mechanism.

Exam practice (5 min) answer: page 49

(a) What is the overall order of the iodine and propanone reaction?

(b) Use the experimental data on page 44 to calculate a value for k.

(c) What are the units of k?

Reaction rates and temperature

Chemical reactions get faster as temperature increases. In many cases the rate of a reaction (and therefore the value of its rate constant, k) approximately doubles with every 10° rise in temperature. This effect of temperature upon reaction rates may be explained by the theory of effective collisions and activation energy.

The jargon

In *colloidal* sulphur the particles are large enough to scatter light (hence the milkiness) but too small to settle as a precipitate.

Checkpoint 1

(a) Describe a simple way of doing this experiment to ensure the same amount of milkiness each time.
(b) Explain how the time interval Δt can be related to the rate of reaction.

Measuring the effect of temperature upon rates

When we add hydrochloric acid to aqueous sodium thiosulphate the clear, colourless solution slowly turns milky white as colloidal sulphur forms. The stoichiometric equation is: $H_2S_2O_3(aq) \rightarrow H_2SO_3(aq) + S(s)$. When we investigate the kinetics we find that the decomposition is a first order reaction and rate = $k_1[H_2S_2O_3(aq)]$. When we perform the same experiment at different temperatures and measure the time interval Δt to the same degree of milkiness, we find Δt approximately halves with each 10° rise in temperature.

The jargon

In the *Arrhenius equation*, A may be called the pre-exponential factor. R is the molar gas constant and T is the absolute temperature (in Kelvin).

Activation energy

If we increase only the temperature and keep all other conditions constant, the reaction rate increases and so the value of the rate constant k must increase.

→ The Arrhenius equation, $k = Ae^{-E_a/RT}$, expresses mathematically the dependence of the rate constant upon the absolute temperature.
→ E_a is called the activation energy of the reaction.

The Arrhenius equation may be written in the form $\ln k = \ln A - E_a/RT$ so that a value for E_a may be determined from the gradient ($= -E_a/R$) of a graph of $\ln k$ against $1/T$.

Watch out!

A and E_a are always positive values. So the power (exponent or index) of e will always be negative ($= -E_a/RT$). The power of e becomes less negative as E_a gets smaller or T gets bigger.

Checkpoint 2

What happens to the value of k (and therefore the reaction rate) if E_a increases and T decreases?

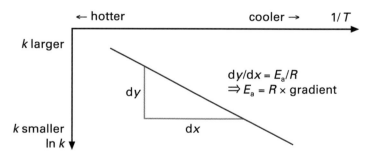

Theory of effective collisions

Billions of close encounters between particles (atoms, ions and molecules) occur every second but only a small number occur as collisions with enough energy (E_a) to activate a reaction.

→ The Arrhenius factor A is a measure the billions of close encounters occurring every second between particles.
→ $e^{-E_a/RT}$ is a measure of the (small) fraction of close encounters that result in collisions causing a reaction.
→ The lower the activation energy and the higher the temperature the faster the reaction.

Grade booster

Even at AS-level you should be able to explain qualitatively why reaction rates increase with temperature and gain marks for your written communication. Not all Awarding Authorities will test you on the Arrhenius equation and its maths. Top grade candidates will understand the equation and the maths.

You may consider activation energy from two points of view.

E_a as the energy barrier

The activation energy may be regarded as the *minimum energy* needed by the reacting particles (atoms, ions or molecules) to achieve a transition state where they exist as an activated complex. You should understand and be able to represent this idea in energy diagrams.

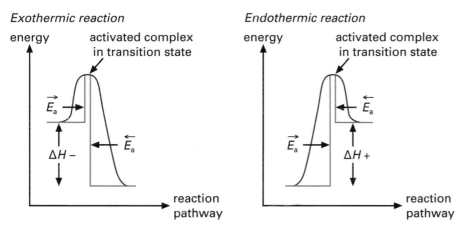

Exothermic reaction

Endothermic reaction

The activation energies appear as barriers that must be surmounted before reactants can turn into products and before products can re-form into reactants.

→ Breaking of strong covalent bonds requires energy and may lead to high activation energies resulting in slow reactions.

E_a and the distribution of energy

As a factor in the Arrhenius equation, E_a determines the proportion of reacting species (atoms, ions or molecules) having enough energy to cross the activation energy barrier. You should understand and be able to represent this idea by sketching energy distribution diagrams. Note that as the temperature rises

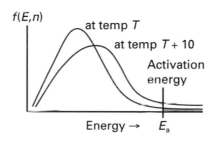

→ the peak of the curve moves *to the right*, so the mean value of $f(E,n)$ and therefore the mean energy of the molecules increases
→ the curve *flattens*, so the total area under it and therefore the total number of molecules remains constant
→ the area under the curve to the right of E_a, and therefore the number of molecules colliding with enough energy to cause a reaction, roughly doubles for every 10° rise in temperature.

The jargon

These energy diagrams may be called *reaction profiles*. The activated complex may be regarded as a transition state in which old bonds have partly broken and new bonds have partly formed.

Checkpoint 3

(a) How is the enthalpy change ΔH for a reaction related to the forward and reverse activation energies?
(b) Explain why the exothermic direction of a reaction is kinetically more feasible than its reverse endothermic direction?

The jargon

In these so-called *Maxwell–Boltzmann distributions of energies*, $f(E,n)$ is a function of energy such that the total area under the curve of $f(E,n)$ against energy is proportional to the total number of reactant species.

Grade booster

Energetically speaking, reactions with very large (positive or negative) values of ΔH may be considered irreversible. Kinetically speaking, reactions with very large values of E_a may be too slow to occur. Do not confuse energetic stability and kinetic stability if you want the top grade.

Exam practice (5 min)   answer: page 49

Sketch a reaction profile for $H_2(g) + I_2(g) \rightarrow 2HI(g)$ and comment on the reversibility of the reaction. [ΔH = –10 kJ mol⁻¹; E_a = 151 kJ mol⁻¹]

Catalysis

So what about those really slow reactions? How can we speed them up? Higher temperature, higher concentrations, but don't forget the catalyst. Homogeneous, heterogeneous and biological: they can all speed up chemical reactions.

The jargon

Negative catalysts or *inhibitors* are substances that slow down chemical processes.

Watch out!

Many heterogeneous catalysts lose efficiency when contaminated with impurities in the gaseous reactants. These catalyst poisons include hydrogen sulphide and carbon monoxide – impurities that must be removed before reaction.

Don't forget!

d-Block transition metal catalysts use their 3d and 4s subshells to adsorb reactant molecules and lower activation energies by weakening bonds in the adsorbed molecules.

Watch out!

Don't confuse adsorb with absorb.

Checkpoint 1

Give the equation for the reaction and name the d-block element used in the heterogeneous catalyst for the industrial production of (a) ammonia in the Haber process and (b) sulphur trioxide in the Contact process.

The jargon

Adsorption is the adhesion of molecules to the surface of a solid or liquid. *Absorption* is the uptake of molecules by a liquid or porous material.

Action point

Make and learn a table of some large-scale industrial processes using catalysts. Include details of the catalyst(s) used and the products being manufactured.

General features of catalysts

→ Catalysts speed up reactions but are not consumed by reactions and therefore do not appear as reactants in the overall equations.
→ Catalysts provide alternative reaction pathways with activation energies lower than those of the uncatalyzed reactions.

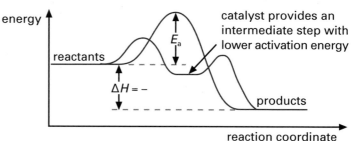

→ Catalysts speed up the rate of attainment of equilibrium for a reversible reaction without altering its composition at equilibrium.
→ Catalysts lower the activation energies of the forward and reverse reactions of a reversible reaction by the same amount.

Heterogeneous catalysis

→ A heterogeneous catalyst is a catalyst in a *different phase* from the reactants.
→ Many industrially important heterogeneous catalysts are d-block transition metals.

Hydrogenation of unsaturated oils to saturated fats in the production of margarine is a simple example. Nickel powder is the catalyst.

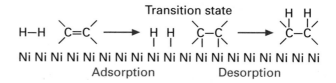

Homogeneous catalysis

→ A homogeneous catalyst is a catalyst in the *same phase* as the reactants.
→ Homogeneous catalysts take part in a reaction so an increase in their concentration will speed up the rate-determining step.

The oxidation of iodide anions by peroxodisulphate(VI) anions is slow: $S_2O_8^{2-}(aq) + 2I^-(aq) \rightarrow 2SO_4^{2-}(aq) + I_2(aq)$. However, iron(II) or iron(III) cations catalyze the reaction by taking part and providing an alternative pathway with a lower overall activation energy.

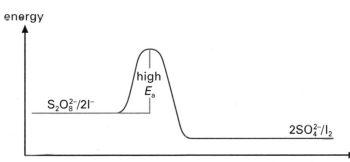

In the catalyzed reaction, iron(II) cations reduce peroxodisulphate anions and iron(III) cations oxidize iodide anions:

$$2Fe^{2+}(aq) + S_2O_8^{2-}(aq) \rightarrow 2Fe^{3+}(aq) + 2SO_4^{2-}(aq)$$
$$2Fe^{3+}(aq) + 2I^-(aq) \rightarrow 2Fe^{2+}(aq) + I_2(aq)$$

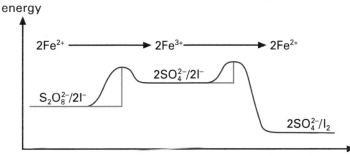

Checkpoint 2

Why might we expect the reaction between two anions to have a high activation energy and be slow?

Examiner's secrets

You will have to produce profiles for reactions that may be exothermic or endothermic and catalyzed or uncatalyzed.

Watch out!

Don't lose marks by labelling the horizontal axis 'time' – a common error. Label it *reaction co-ordinate, reaction pathway* or *extent of reaction*.

Autocatalysis

In most reactions we expect the concentration of catalyst to remain constant. In the iodination of propanone, the concentration of the homogeneous catalyst, $H^+(aq)$, actually increases because hydriodic acid is one of the products. Consequently, the rate of the reaction at first increases with time even though the concentration of propanone is decreasing.

→ A reaction is autocatalytic if it is catalyzed by one of its products.

The jargon

Enzymes are biochemical catalysts that usually work best around 36 °C and at pH around 7. At high temperatures and extreme pH values the protein structure is disrupted and the active sites of the substrate are lost.

Exam practice (10 min)   answers: page 50

1 (a) Suggest why E_a for the reaction of Fe^{2+} with the peroxodisulphate ion might be larger than E_a for the reaction of Fe^{3+} with the iodide ion.
   (b) State the effect, if any, of the catalyst upon the value of ΔH.
   (c) Redraw the catalyzed reaction profile to show iron(III) cations acting as the catalyst.

2 The ethanedioate ion reacts with the manganate(VII) ion in excess aqueous acid to form carbon dioxide and the manganese(II) ion. Write a balanced ionic equation for the redox and suggest *two* reasons why the reaction accelerates initially.

Action point

Make and learn a list of catalysts used in organic chemistry and include examples of reactions in which they are used. This knowledge will help you gain the top grade.

Structured exam question

answers: page 50

(a) The diagram below shows the distribution of molecular velocities in a given mass of gas at a particular temperature. The vertical axis is a function of the velocity.

Draw on the grid a curve showing the distribution of molecular velocities
  (i) at a lower temperature – label this curve L
  (ii) at a higher temperature – label this curve H

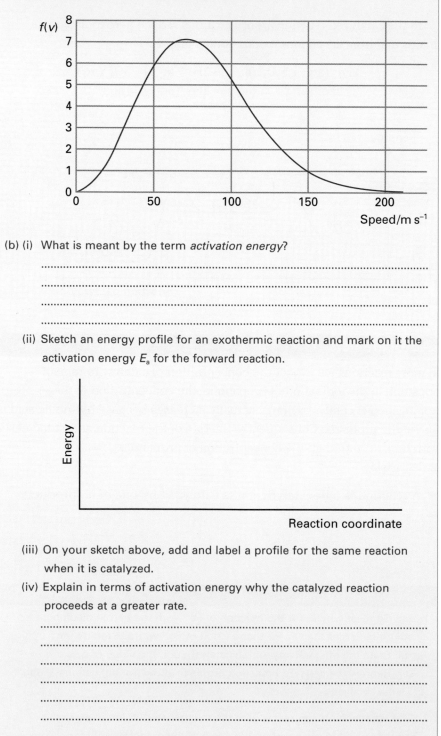

(b) (i) What is meant by the term *activation energy*?

....
....
....
....

(ii) Sketch an energy profile for an exothermic reaction and mark on it the activation energy E_a for the forward reaction.

(iii) On your sketch above, add and label a profile for the same reaction when it is catalyzed.

(iv) Explain in terms of activation energy why the catalyzed reaction proceeds at a greater rate.

....
....
....
....
....
....
....
....

(10 min)

Answers
Kinetics

Rates and orders of reaction

Checkpoints

1 (a) Magnesium metal must be clean. Any oxide coating would react first to give water instead of hydrogen gas. Care must be taken that the syringe does not stick. The volume measured would be incorrect. The amount of reactants must not produce a volume of hydrogen greater than the capacity of the syringe.

(b) Put the acid in a test tube inside the conical flask which can be tilted to allow the acid to make contact with the metal as the clock is started.

2 (a) To prevent loss of mass due to spray from the acid.

(b) $\Delta m = (m_t - m_f) \propto [HCl(aq)]_t$, the concentration of unreacted hydrochloric acid at time t.

Exam practice

(a) $^{14}_{6}C \rightarrow {}^{14}_{7}N + \beta$

(b) It would take three half-lives, i.e. 17 190 years.

(c) The isotope is the basis for carbon dating of relics.

Orders of reaction and mechanisms

Checkpoints

1 Colorimetry, because the intensity of the brown colour of aqueous iodine decreases with time as the $I_2(aq)$ is converted into colourless $I^-(aq)$.
Sampling, because the unreacted iodine could be titrated with aqueous sodium thiosulphate using starch as an indicator for the end-point of the titration.

2 (a)   Rate = $k[N_2O_5]$

(b)   First order (1)

(c)

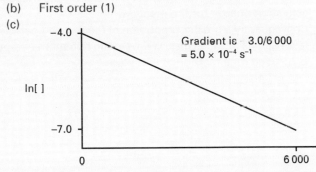

Gradient is $-3.0/6\,000$
$= 5.0 \times 10^{-4}\,s^{-1}$

Rate constant $k = 5.0 \times 10^{-4}\,s^{-1}$

3 $(CH_3)_2CO(aq) + I_2(aq) \rightarrow CH_3COCH_2I(aq) + HI(aq)$

Exam practice

(a) The overall order is 2.

(b) According to the rate equation

$$rate = k[CH_3COCH_3][HCl]$$

$$k = \frac{rate}{[CH_3COCH_3][HCl]}$$

$$k = \frac{1.1 \times 10^{-5}\,mol\,dm^{-3}}{0.5\,mol\,dm^{-3} \times 1.25\,mol\,dm^{-3}}$$

$k = 2 \times 10^{-5}\,mol\,dm^{-3}$ (only 1 significant figure since data to only one significant figure).

(c) $mol^{-1}\,dm^3\,s^{-1}$.

Reaction rates and temperature

Checkpoints

1 (a) Draw a black cross on a piece of paper. Put the paper under the reaction vessel and look at the cross through the reaction mixture. Record the time taken for the milkiness to obscure the cross from view.

(b)

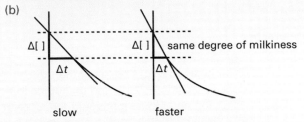

slow   faster

From a graph of concentration [ ] against time t, rate $\propto \Delta[\ ]/\Delta t$ but $\Delta[\ ]$ is constant, so rate $\propto 1/\Delta t$. The higher the rate, the smaller Δt.

2 If E_a increases k will decrease.
If temperature decreases k will decrease.

3 (a) $\Delta H = E_{a(forward)} - E_{a(reverse)}$
Note that this expression always gives the correct sign for ΔH because activation energy values are always positive.

(b) For an exothermic reaction the energy of the products is lower than the energy of the reactants and so the activation energy of the reverse reaction must be greater than the activation energy of the forward reaction. Thus the forward (exothermic) reaction is kinetically more likely to occur than the reverse (endothermic) reaction.

Exam practice

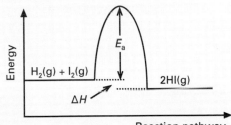

The difference between the activation energy of the forward and reverse reactions is very small, so the reaction should be reversible. The activation energies are high, so the reaction should be slow to attain equilibrium.

Catalysis

Checkpoints

1 (a) $N_2(g) + 3H_2(g) \rightleftharpoons 2NH_3(g)$; iron
   (b) $2SO_2(g) + O_2(g) \rightleftharpoons 2SO_3(g)$; vanadium(V) oxide
2 The activation energy will be high since two negatively charged particles have to approach each other and come together. The like charges will repel and tend to keep them apart.

Exam practice

1 (a) The charge density of the Fe^{2+} ion and of the peroxodisulphate ion might be lower than the charge densities of the Fe^{3+} ion and the iodide ion. If so, the Fe^{3+} ion and iodide ion would attract each other more strongly than would the Fe^{2+} ion and peroxodisulphate ion. This might make the reaction between Fe^{3+} ions and iodide ions faster than the reaction between Fe^{2+} ions and peroxodisulphate ions. However, both reactions will have low activation energies and be fast.
   (b) No effect since we have the same reactant and the same products.
   (c)

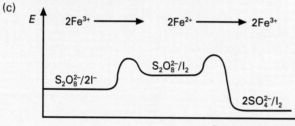

2 $5C_2O_4^{2-}(aq) + 16H^+(aq) + 2MnO_4^-(aq)$
   $\rightarrow 10CO_2(g) + 2Mn^{2+}(aq) + 8H_2O(l)$
   A product, Mn^{2+} ions, catalyzes the reaction.
   Heat from the redox reaction raises the temperature.

Structured exam question

(a)

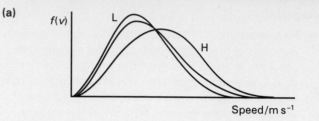

Notice that the distribution curve gets lower and fatter as the temperature rises *but* the area under the curve does *not* change because it is proportional to the total number of molecules.

(b) (i) The activation energy E_a is the minimum energy needed by reacting particles (atoms, ions or molecules) to achieve the transition state so that a reaction may occur between them.

(b) (ii), (iii)

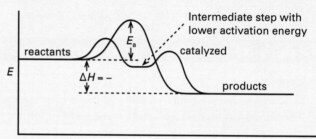

(iv) A catalyst provides an alternative route for the reaction. This has the net effect of lowering the overall activation energy for the reaction.
Consequently a greater proportion of the reacting particles (atoms, ions or molecules) will have the minimum energy, E_a, needed for a reaction to take place.

Equilibria

You read and write equations from left to right in the direction → of the chemical change or forward reaction. You may know that some chemical changes can also go backwards ← or in the reverse direction. In this chapter you will learn about reversible reactions ⇌ which include mixtures of gases and, more importantly, acids and bases in aqueous solution. You must be able to apply the law of chemical equilibrium to these reactions. First make sure you understand the principles behind the law before you worry about the maths involved.

By the end of this chapter you will be able to

→ Describe and explain dynamic equilibria and apply Le Chatelier's principle to reactions at equilibrium

→ Write expressions for equilibrium constants K_c and K_p, and using supplied data, calculate values for K_c or work out equilibrium concentrations of reactants and products

→ Calculate partial pressures of gases in order to calculate K_p, and given a value of K_p work out partial pressures of products or reactants at equilibrium

→ Define and give examples of strong and weak acids and bases and give a Brønsted–Lowry definition of acid and base

→ Recognize conjugate acids and bases and work out pH of solutions from concentrations of $H^+(aq)$ or $OH^-(aq)$

→ Give expressions for K_a and K_b and use supplied values of K_a and K_b (or pK_a and pK_b) to work out the pH of solutions

→ Sketch pH curves and interpret ones supplied to choose a suitable indicator for a titration

→ Define a buffer, describe how it works and calculate its pH

Topic checklist

Tick each of the boxes below when you are satisfied that you have mastered the topic

	Edexcel		AQA		OCR		WJEC		CCEA	
	AS	A2	AS	A2	AS	A2	AS	A2	AS	A2
Dynamic equilibria		●		●	○		○		○	
Chemical equilibria		●		●		●		●		●
Equilibrium problems	○	●	○	●	○	●	○	●	○	●
Aqueous equilibria		●		●		●		●		●
Weak acids and bases		●		●		●		●		●
Acid–base reactions		●		●		●		●		●
Buffers and indicators		●		●		●		●		●

Dynamic equilibria

Reactions usually carry on until all of the reactants have turned into products or until one or other of the reactants runs out, but this isn't always the case. Take, for example, the reaction between nitrogen and hydrogen to make ammonia. As the ammonia builds up it starts to decompose back to nitrogen and hydrogen. Eventually, the reaction seems to stop with only about 10% of the elements converted into ammonia under normal conditions. The forward and reverse reactions are occurring at the same rate and a dynamic equilibrium is reached.

Phase equilibria

The melting point and boiling point of elements and compounds are two characteristic properties listed in data books.

→ The melting point of a substance is the temperature at which the solid and liquid phases can coexist in equilibrium (at 1 atm).

When a solid gains energy, its particles (atoms, ions or molecules) vibrate more vigorously about their fixed positions in the lattice. When a solid gains enough energy to melt, its particles leave the lattice and their kinetic energy changes from vibrational energy into translational and rotational energy. When a liquid (or solid) evaporates, its particles escape from the surface to form a vapour.

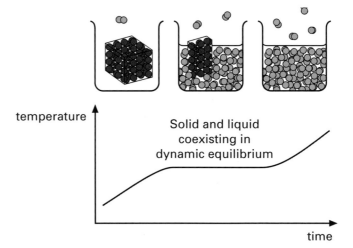

The converse is true when a vapour condenses and a liquid solidifies.

In an open vessel, liquids evaporate as their particles escape into the surrounding air. In a closed vessel at a fixed temperature, liquid and vapour attain dynamic equilibrium. The rate at which the particles escape from the liquid equals the rate at which they return from the vapour.

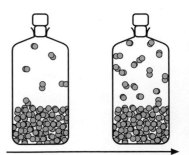

Increasing temperature

→ The vapour pressure of a liquid increases with increasing temperature.

The jargon

Strictly speaking, a *gas* is above its critical temperature and unable to condense into a liquid phase. A *vapour* is below its critical temperature and able to condense into a liquid with a visible surface.

Checkpoint 1

(a) What is the connection between melting point and freezing point?
(b) Explain why ice (i) will cool a drink and (ii) will keep it cool.
(c) Suggest why at room temperature moth balls (naphthalene, $C_{10}H_8$) give off a vapour but salt (NaCl) does not.

Watch out!

Evaporation is an endothermic process. A liquid in an open container will take heat from the surroundings and evaporate completely because its vapour molecules escape into the air and equilibrium is never attained.

Links

See page 18: intermolecular forces.

→ The boiling point of a substance is the temperature at which the vapour pressure of the liquid equals 1 atm (101 325 Pa) or the external pressure when it is not 1 atm.

Chemical equilibria

Reversible reactions are chemical changes that

→ may take place in either direction depending upon the conditions
→ may be homogeneous (one phase) or heterogeneous (two or more phases)
→ may attain dynamic equilibrium in which the rate of the forward reaction equals the rate of the reverse reaction

You will meet a variety of reversible reactions including the following:

$$CH_3CO_2C_2H_5(l) + H_2O(l) \rightleftharpoons CH_3CO_2H(l) + C_2H_5OH(l)$$
$$N_2O_4(g) \rightleftharpoons 2NO_2(g)$$
$$N_2(g) + 3H_2(g) \rightleftharpoons 2NH_3(g); \Delta H^{\ominus} = -94.4 \, kJ \, mol^{-1}$$
$$2SO_2(g) + O_2(g) \rightleftharpoons 2SO_3(g); \Delta H^{\ominus} = -197 \, kJ \, mol^{-1}$$

Law of chemical equilibrium

In general, a reversible reaction can be represented by the equation

$$aA + bB \rightleftharpoons cC + dD$$

At equilibrium at a constant temperature T, then $[C]^c \times [D]^d$ divided by $[A]^a \times [B]^b$ has a constant value, K_c, if the chemicals are all aqueous or in the liquid state. If the chemicals are in the gaseous state, then

$$\frac{p_C^c \times p_D^d}{p_A^a \times p_B^b} = K_p \quad \text{where } p \text{ is the partial pressure}$$

The letter K stands for the value of the constants, usually called equilibrium constants. If we use a subscript (eq) to show that the concentration or partial pressure values are the values at equilibrium, then we can write

$$K_c = \frac{[C]_{eq}^c \times [D]_{eq}^d}{[A]_{eq}^a \times [B]_{eq}^b} \quad \text{or} \quad K_p = \frac{p_{C,eq}^c \times p_{D,eq}^d}{p_{A,eq}^a \times p_{B,eq}^b}$$

The actual value and units of the equilibrium constants, K_c and K_p, depend on the balanced equation for the reversible reaction. The value of K_c or K_p for a reaction is the reciprocal of K_c and K_p for its reverse reaction and the units alter accordingly.

Watch out!

Do not use [ ] when writing expressions for K_p.

Links

See page 58: aqueous equilibria; page 127: ligand replacement reactions of complex ions.

Checkpoint 2

What would be the units of k_c and K_p if (i) $a + b = c + d$, (ii) $a = 1$, $b = 3$, $c = 2$, $d = 0$?

Don't forget!

When writing the expression for K for a reaction put **Products** on the top and **Reactants** underneath.
$\frac{[C]^c \times [D]^d}{[A]^a \times [B]^b}$ could have any value from $-\infty$ to 0 to $+\infty$ if the reaction is NOT at equilibrium but for a specific reaction at a specific temperature it can have only one value which we represent by K_c.

Exam practice   5 minutes   answers: page 67

(a) What is meant by 'standard molar enthalpy change of vaporization'?

(b) Suggest why the value for water is almost seven times that for hexane.

(c) Explain why water boils below 100 °C at a high altitude in the Himalayas.

Chemical equilibria

So what do K_c and K_p tell us and how can we change them? The bigger K_c and K_p are, the higher the percentage conversion of reactants and the more products we expect to get. For a specific reaction only one factor can change the size of K and that is the temperature; in this section we show you how.

Factors affecting equilibrium constant values

Nature of the reversible reaction

The value represented by K_c and K_p for a given temperature depends upon the chemical system being considered.

Chemical system at 500 K	K_p	Units	$\Delta H^{\ominus}_{500}/\text{kJ mol}^{-1}$
$H_2(g) + CO_2(g) \rightleftharpoons H_2O(g) + CO(g)$	7.8×10^{-3}	–	$+41$
$N_2(g) + 3H_2(g) \rightleftharpoons 2NH_3(g)$	3.8×10^{-2}	atm^{-2}	-101
$H_2(g) + I_2(g) \rightleftharpoons 2HI(g)$	25.0	–	-10
$N_2O_4(g) \rightleftharpoons 2NO_2(g)$	1.7×10^3	atm	$+57$
$2SO_2(g) + O_2(g) \rightleftharpoons 2SO_3(g)$	2.5×10^{10}	atm^{-1}	-200

From the above K_p values we can tell that at 230 °C very little reaction occurs between hydrogen and carbon dioxide or nitrogen, whereas the dissociation of dinitrogen tetraoxide is considerable and the formation of sulphur trioxide is almost complete.

The chemical equation

The value represented by K_c and K_p for a given temperature depends upon the way we write the equation for the reaction.

T/°C	Balanced equation	K_p	Units	K_c	Units
80	$N_2O_4(g) \rightleftharpoons 2NO_2(g)$	4	atm	0.138	mol dm^{-3}
80	$\frac{1}{2}N_2O_4(g) \rightleftharpoons NO_2(g)$	2	$\text{atm}^{1/2}$	0.371	$\text{mol}^{1/2}\,\text{dm}^{-3/2}$
80	$NO_2(g) \rightleftharpoons \frac{1}{2}N_2O_4(g)$	$\frac{1}{2}$	$\text{atm}^{-1/2}$	2.70	$\text{mol}^{-1/2}\,\text{dm}^{3/2}$
80	$2NO_2(g) \rightleftharpoons N_2O_4(g)$	$\frac{1}{4}$	atm^{-1}	7.25	$\text{mol}^{-1}\,\text{dm}^3$

Temperature of the equilibrium system

The value represented by K_c and K_p for a given chemical system depends upon the temperature at which the system is being kept.

For the dissociation $N_2O_4(g) \rightleftharpoons 2NO_2(g)$; $\Delta H^{\ominus} = +58 \text{ kJ mol}^{-1}$,

K_p/atm:	0.115	3.89	47.9	347	1 700	6 030
T/°C:	25	77	127	177	227	277

If we warm an equilibrium mixture of dinitrogen tetraoxide and nitrogen dioxide (NO_2, brown), we see the colour become darker brown as the K_p value increases. If we cool the mixture, we see the colour become lighter brown as the K_p value decreases.

→ K_c and K_p for the *endothermic* direction of a reaction will *increase* with increasing temperature.
→ K_c and K_p for the *exothermic* direction of a reaction will *decrease* with increasing temperature.

Total pressure

→ If the total number of moles of gaseous products is greater than that of the gaseous reactants, an increase in the total pressure is accompanied by a shift in the composition in favour of reactants.

Composition of equilibrium mixture of $N_2O_4(g)$ and $NO_2(g)$ at 80 °C

$p\,N_2O_4$/atm	$p\,NO_2$/atm	Total pressure/atm	mol % NO_2	mol % N_2O_4
0.17	0.83	1	83	17
0.54	1.46	2	73	27
1.00	2.00	3	67	33
1.53	2.47	4	62	38

→ If the total number of moles of gaseous products is the same as that of the gaseous reactants, the equilibrium composition does not change when the total pressure changes.

Partial pressure (or concentration)

→ If the partial pressure (or concentration) of a component is altered, the equilibrium may be disturbed and the composition may alter to restore equilibrium.

The following data show the result of adding extra iodine to an equilibrium mixture of iodine, hydrogen and hydrogen iodide.

Composition of equilibrium mixture of $H_2(g)$, $I_2(g)$ and $HI(g)$ at 227 °C

	Amount I_2 added/mol	$I_2(g)$/mol	$H_2(g)$/mol	$HI(g)$/mol	K
(a)	0	1.000	1.000	5.000	25.00
(b)	1	1.720	0.719 5	5.561	25.00
(c)	2	2.548	0.547 5	5.905	25.00
(d)	4	4.361	0.361 4	6.277	25.00

Add 1 mol I_2 to 1 mol I_2 in a flask on its own and you have 2 mol I_2 in the flask. Add 1 mol I_2 to 1 mol I_2 in equilibrium with 1 mol H_2 and 5 mol HI and you have less than 2 mol I_2 in the flask. Why? Because some of the added iodine combines with (and decreases) the hydrogen present to produce extra hydrogen iodide.

→ If in a reversible reaction at equilibrium the partial pressure (or concentration) of one reactant is increased, the partial pressures of the other reactants decrease and the partial pressures of the products increase to restore the chemical equilibrium.

Checkpoint 1

Use the law of chemical equilibrium to explain why the industrial synthesis of ammonia from nitrogen and hydrogen by the Haber process is carried out at a total pressure of about 200 atm.

Watch out!

The value of K_p for the dissociation of $N_2O_4(g)$ at 80 °C is constant at 4 atm for all these equilibrium mixtures.

Checkpoint 2

Use the law of chemical equilibrium to explain why, in the Contact process for the industrial production of sulphuric acid, sulphur dioxide is reacted with an excess of air.

Links

See page 21: partial pressure.

Don't forget!

The principle put forward in 1888 by Henri Le Chatelier describes *qualitatively* the effects of stress upon a system in equilibrium. The law of chemical equilibrium describes the effects *quantitatively*.

Exam practice   6 minutes   answers: page 67

State Le Chatelier's principle. Write an expression for K_p for the endothermic dissociation of $PCl_5(g)$ into $PCl_3(g)$ and $Cl_2(g)$. Explain the effect upon the dissociation of increasing (i) the total pressure and (ii) the temperature.

Equilibrium problems

You should be able to predict and explain changes in chemical equilibria with temperature, pressure or concentration. You should also be able to solve simple problems involving K_c, K_p and equilibrium composition.

Le Chatelier's principle

The law of chemical equilibrium deals *quantitatively* with the way the composition of an equilibrium system may change with temperature, pressure and concentration. In 1888 Henri Le Chatelier put forward his principle which dealt *qualitatively* with the effects of stress upon a system in equilibrium. There are probably as many different English translations as there are textbooks.

→ If you alter the conditions of a reversible reaction and disturb the equilibrium, the composition of the mixture may change to restore the equilibrium and to minimize the effect of altering the conditions.

According to Le Chatelier's principle, the yield of sulphur trioxide in the Contact process would be increased by increasing the total pressure because the formation of $SO_3(g)$ is accompanied by a decrease in volume (3 volumes to 2 volumes): $2SO_2(g) + O_2(g) \rightleftharpoons 2SO_3(g)$.

Problems

Calculating a value for K_c or K_p

Example: The degree of dissociation of dinitrogen tetraoxide into nitrogen dioxide is 0.5 at 60 °C and a total pressure of 1 atm: $N_2O_4(g) \rightleftharpoons 2NO_2(g)$. Calculate the equilibrium constant K_p at this temperature.

$$N_2O_4(g) \rightarrow 2NO_2(g)$$
$$1 \text{ mol} \rightarrow 2 \text{ mol}$$
so
$$0.5 \text{ mol} \rightarrow 1 \text{ mol}$$

If 0.5 of 1 mol N_2O_4 dissociates, the resulting gas mixture will contain 0.5 mol (undissociated) N_2O_4 and 1 mol NO_2, so the amounts of the two gases will be in the ratio $1:2$.

$$\text{partial pressure } N_2O_4 = {}^1\!/_3 \text{ atm}; NO_2 = {}^2\!/_3 \text{ atm}$$

$$K_p = \frac{p^2_{NO_2}}{p_{N_2O_4}} = \frac{({}^2\!/_3)^2}{{}^1\!/_3}$$

Hence $K_p = 1^1\!/_3$ atm.

Calculating the composition of a mixture

Example: At a temperature of 650 °C and a total pressure of 3 atm, dimeric aluminium chloride dissociate into monomers according to the equation $Al_2Cl_6(g) \rightleftharpoons 2AlCl_3(g)$. The value of K_p is 4.0 atm at 650 °C. Calculate the percentage dissociation of the aluminium chloride.

$$Al_2Cl_6(g) \rightarrow 2AlCl_3(g)$$
$$1 \text{ mol} \rightarrow 2 \text{ mol}$$

If x mol dissociates then the composition of the mixture will be

$$(1 - x) \text{ mol} + 2x \text{ mol} = (1 + x) \text{ mol total}$$

and the partial pressures of the components will be

$$p_d = 3 \times (1 - x)/(1 + x) \text{ and } p_m = 3 \times 2x/(1 + x)$$

K_p is

$$(p_m)^2/p_d = 4.0 = [3 \times 2x/(1 + x)]^2/[3 \times (1 - x)/(1 + x)]$$
$$4.0 \times [3 \times (1 - x)/(1 + x)] = [3 \times 2x/(1 + x)]^2$$
$$4.0 \times 3 \times (1 - x) \times (1 + x) = (3 \times 2x)^2$$
$$12.0 \times (1 - x^2) = 36x^2 \qquad 12.0 - 12.0x^2 = 36x^2$$
$$12.0 = 48x^2 \qquad 1.0 = 4x^2$$

which simplifies to

$$x = {}^1\!/_2$$

Aluminium chloride is 50% dissociated.

Example: Water and carbon monoxide in the molar ratio of 1 : 1 are reacted together at 600 °C. The equation for the reaction is
$H_2O(g) + CO(g) \rightleftharpoons H_2(g) + CO_2(g)$
and the value of the equilibrium constant is 2.9 at 600 °C.
Calculate the mole percentage of hydrogen in the mixture at equilibrium.

$$H_2O(g) + CO(g) \rightarrow H_2(g) + CO_2(g)$$
$$1 \text{ mol} + 1 \text{ mol} \rightarrow 1 \text{ mol} + 1 \text{ mol}$$

Let V dm^3 equilibrium mixture have x mol H_2. Then the composition of the mixture will be

$$(1 - x) + (1 - x) \rightarrow x + x$$

and the concentrations of the components will be

$$(1 - x)/V + (1 - x)/V \rightarrow x/V + x/V$$

So

$$\frac{(x/V)(x/V)}{[(1 - x)/V][(1 - x)/V]} = 2.9$$

which simplifies to

$$x^2/(1 - x)^2 = 2.9$$

Take the square root of both sides

$$x/(1 - x) = +1.70 \text{ (why reject } -1.70 ?)$$
$$x = 0.63$$
$$H_2O(g) + CO(g) \rightarrow H_2(g) + CO_2(g)$$
$$0.37 \qquad 0.37 \qquad 0.63 \qquad 0.63 \quad = 2 \text{ mol total}$$

0.63 mol H_2 in 2 mol mixture, 31.5% by mole of hydrogen at equilibrium.

Watch out!

This first question is easier than the next one because you do not need to solve a quadratic equation.

Examiner's secrets

You could omit this working from your answer. These lines are included in case you would like some help with the maths.

Grade booster

A-level questions don't come much harder than this but you could still score quite a few marks even if you do not solve the equation. This is the kind of working you should show.
'Which simplifies to' means the Vs cancel to give x^2 at the top (numerator) and at the bottom (denominator) the expression $(1 - x)(1 - x)$.
It is unlikely that you will be given examples which need you to solve a quadratic equation.
You should spot that you can take the square root of both sides. So $x/(1 - x) - \sqrt{(2.9)}$. This is a simple equation in x. If, say, the mole ratio 3 : 1 for H_2O and CO had been used then you would have to have solved a quadratic equation.

Exam practice   5 minutes   answers: page 67

2 mol ethanoic acid $CH_3CO_2H(l)$ and 1 mol ethanol $C_2H_5OH(l)$ react to form an equilibrium mixture containing 0.85 mol ethyl ethanoate at 60 °C. Write an equation for the esterification reaction and calculate the value of K_c.

Aqueous equilibria

The law of chemical equilibrium applies to the ionization of water and to aqueous acid–base reactions. K_w, K_a and K_b are three important constants.

Ionization of water

The fundamental equilibrium that exists in water and all aqueous solutions is $H_2O(l) + H_2O(l) \rightleftharpoons H_3O^+(aq) + OH^-(aq)$. When we apply the law of chemical equilibrium (and some simplifying mathematical assumptions) to this proton-transfer reaction we can write

$$[H_3O^+(aq)][OH^-(aq)] = K_w$$
$$K_w = 1 \times 10^{-14} \text{ (or 0.000 000 000 000 01) mol}^2\text{dm}^{-6} \text{ at } 25\,°C$$

Even expressed in standard form as a negative power of 10, the value of K_w is an inconvenient number. So we write

$$pK_w = -\log_{10}(K_w/\text{mol}^2\text{dm}^{-6}) = 14 \text{ at } 25\,°C$$

In pure water the concentration of $H_3O^+(aq)$ must equal the concentration of $OH^-(aq)$. So it follows that

$$[H_3O^+(aq)] = [OH^-(aq)] = 1 \times 10^{-7} \text{ mol dm}^{-3} \text{ at } 25\,°C$$

For convenience, we write

$$pH = -\log_{10}([H_3O^+(aq)]/\text{mol dm}^{-3})$$
$$pOH = -\log_{10}([OH^-(aq)]/\text{mol dm}^{-3})$$

These two definitions combine with the definition of pK_w to give the following important relationship:

→ $pK_w = pH + pOH = 14 \text{ at } 25\,°C$

pH scale of acidity

An aqueous solution is neutral if $pH = pOH = 7$ at $25\,°C$.

Acid solutions

Hydrogen chloride gas dissolves rapidly in water and ionizes completely: $HCl(g) + H_2O(l) \rightarrow H_3O^+(aq) + Cl^-(aq)$.
Consequently, in a 0.1 mol dm^{-3} solution of hydrochloric acid,

$$[H_3O^+(aq)] = 1 \times 10^{-1} \text{ mol dm}^{-3}$$

Therefore $pH = -\log_{10}([H_3O^+(aq)]/\text{mol dm}^{-3}) = 1$. But $pH + pOH = 14$ at $25\,°C$. Therefore $pOH = 13$.

→ An aqueous solution is acidic if $pH < 7$.

Alkaline solutions

Sodium hydroxide pellets dissolve rapidly in water to give aqueous sodium ions and hydroxide ions: $NaOH(s) \rightarrow Na^+(aq) + OH^-(aq)$. Consequently, in a 0.1 mol dm^{-3} solution of sodium hydroxide,

$$[OH^-(aq)] = 1 \times 10^{-1} \text{ mol dm}^{-3}$$

Therefore $pOH = -\log_{10}([OH^-(aq)]/\text{mol dm}^{-3}) = 1$. But $pH + pOH = 14$ at $25\,°C$. Therefore $pH = 13$.

→ An aqueous solution is alkaline if $pH > 7$.

The jargon

You can use either $H^+(aq)$ or $H_3O^+(aq)$ when discussing *hydrogen ion concentration*.

Checkpoint 1

How might the value of K_w change with increasing temperature?

Grade booster

If you want the top grade learn this definition of **pH**. It is very important and usually tested.

Don't forget!

At A-level you may assume that the temperature is $25\,°C$ unless you are told otherwise. Consequently, you may say that $pH = 7$ means a neutral solution.

Checkpoint 2

Calculate the pH of an acidic solution whose $[H_3O^+(aq)] = 2 \times 10^{-2}$ mol dm^{-3}.

Checkpoint 3

Calculate the pOH of a basic solution whose $[OH^-(aq)] = 2 \times 10^{-2}$ mol dm^{-3}. What would be its pH?

pH range

In principle we could have acidic solutions with pH < 0 and basic solutions with pOH < 0. In practice we work mostly with solutions in the pH range from 1 to 13.

$$[H_3O^+(aq)] \rightarrow 10^{-1} \qquad\qquad 10^{-7} \qquad\qquad 10^{-13}$$
$$/mol\,dm^{-3} \quad |\qquad\qquad\quad | \qquad\qquad\quad |$$

| pH | 1 | 2 | 3 | 4 | 5 | 6 | 7 | 8 | 9 | 10 | 11 | 12 | 13 |

| acidic | neutral | alkaline |
| $[H^+] > [OH^-]$ | $[H^+] = [OH^-]$ | $[H^+] < [OH^-]$ |

| 13 | 12 | 11 | 10 | 9 | 8 | 7 | 6 | 5 | 4 | 3 | 2 | 1 | **pOH** |

$$| \qquad\qquad\qquad | \qquad\qquad\qquad | \quad [OH^-(aq)]$$
$$10^{-13} \qquad\qquad 10^{-7} \qquad\qquad 10^{-1} \leftarrow /mol\,dm^{-3}$$

Strong acids and bases

Mineral acids such as hydrochloric, nitric and sulphuric acid are strong acids and the s-block metal hydroxides such as sodium, potassium, calcium and barium hydroxide are strong alkalis.

→ Strong acids and strong bases are completely ionized in water.

Calculations

You could be asked to do simple calculations involving pH and the composition of aqueous strong acids and bases. The problems fall into two types. Here is an example of each type.

Example 1: What is the concentration of dilute nitric acid of pH = 1.4?

$$pH = 1.4 = -\log_{10}([H_3O^+(aq)]/mol\,dm^{-3})$$
$$[H_3O^+(aq)] = 4 \times 10^{-2}\,mol\,dm^{-3}$$

but nitric acid is strong: $HNO_3(aq) \rightarrow H^+(aq) + NO_3^-(aq)$. Therefore the concentration of the nitric acid is $4 \times 10^{-2}\,mol\,dm^{-3}$ or $0.04\,mol\,dm^{-3}\,HNO_3(aq)$.

Example 2: What is the pH of $0.2\,mol\,dm^{-3}\,Ba(OH)_2(aq)$?

$$0.02\,mol\,Ba(OH)_2(aq) \rightarrow 0.02\,mol\,Ba^{2+} \text{ but } 0.04\,mol\,OH^-$$
$$[OH^-(aq)] = 4 \times 10^{-2}\,mol\,dm^{-3}$$

but

$$pOH = -\log_{10}([OH^-(aq)]/mol\,dm^{-3})$$
$$pOH = 1.4$$

but

$$pOH + pH = 14$$
$$pH = 12.6$$

Exam practice (5 minutes)   answers: page 68

(a) Define pH.

(b) Calculate the approximate pH of aqueous sulphuric acid of concentration $0.05\,mol\,dm^{-3}$ stating clearly any assumptions made.

(c) Calculate the concentration, in $g\,dm^{-3}$, of calcium hydroxide in limewater whose pH is 10.8.

Watch out!

$Ba(OH)_2(s)$ is an ionic solid 0.1 mol of which gives 0.2 mol OH (aq) when it is dissolved in water.
$H_2SO_4(l)$ is a covalent liquid 0.1 mol of which gives $0.2\,mol\,H_3O^+(aq)$ when it reacts with water:
$H_2SO_4(aq) \rightarrow 2H^+(aq) + SO_4(aq)$

The jargon

We may simplify an equation like
$HNO_3(aq) + H_2O(l) \rightarrow H_3O^+(aq) + NO_3^-(aq)$
by writing
$HNO_3(aq) \rightarrow H^+(aq) + NO_3^-(aq)$

Grade booster

You will be given simple numbers and easy arithmetic where possible. When you use your calculator for the logs (base 10 remember) round your answer to an appropriate significant figure or you may lose marks and the top grade!

The jargon

Limewater is a saturated aqueous solution of calcium hydroxide.

Checkpoint 4

What gas is identified using limewater and how does the test work?

Weak acids and bases

The law of chemical equilibrium applies to the partial ionization of weak acids and bases. The dissociation constants, K_a and K_b, are important constants.

Dissociation of weak acids and bases

Weak acids and bases are only partially ionized in water.

Weak acids

Liquid ethanoic acid mixes in all proportions with water but fewer than 10% of its molecules ionize into hydrogen ions and ethanoate ions:

$$CH_3CO_2H(aq) + H_2O(l) \rightleftharpoons H_3O^+(aq) + CH_3CO_2^-(aq)$$

When we apply the law of chemical equilibrium (and some simplifying mathematical assumptions) to this proton-transfer reaction we can write

$$\frac{[H_3O^+(aq)][CH_3CO_2^-(aq)]}{[CH_3CO_2H(aq)]} = K_a \text{ (the acid dissociation constant)}$$

At 25 °C the value of K_a for ethanoic acid is 1.7×10^{-5} mol dm^{-3}. We can obtain a more convenient number by defining

➡ $pK_a = -\log_{10}(K_a/\text{mol dm}^{-3})$
➡ pK_a for ethanoic acid = 4.8 and is typical for carboxylic acids

Weak bases

Ammonia gas dissolves rapidly in water but less than 10% of the dissolved ammonia ionizes into ammonium ions and hydroxide ions:

$$NH_3(aq) + H_2O(l) \rightleftharpoons NH_4^+(aq) + OH^-(aq)$$

When we apply the law of chemical equilibrium (and some simplifying mathematical assumptions) to this proton-transfer reaction we can write

$$\frac{[NH_4^+(aq)][OH^-(aq)]}{[NH_3(aq)]} = K_b \text{ (the base dissociation constant)}$$

At 25 °C the value of K_b for aqueous ammonia is 1.8×10^{-5} mol dm^{-3}. We can obtain a more convenient number by defining

➡ $pK_b = -\log_{10}(K_b/\text{mol dm}^{-3})$
➡ pK_b for aqueous ammonia = 4.8 and is similar for alkyl amines

Calculations

You could be asked to do calculations involving pH, K_a, K_b and the composition of a solution. At A-level you can usually make the following simplifying assumptions.

1 The weak acid (or alkali) supplies all the H$_3$O$^+$(aq) ions (or all the OH$^-$(aq) ions) for the solution, the amount coming from the water being negligible.
2 The weak acid (or alkali) ionizes so little that it can be regarded as being un-ionized when calculating the concentration of its (undissociated) molecules.

There are three types of problem.

1 Calculate pH from the concentration and K_a (or K_b) of an aqueous weak acid (or alkali). This is very popular in examinations.
2 Calculate K_a (or K_b) of a weak acid (or alkali) from the concentration and pH of its solution. This occurs in connection with titration curves.
3 Calculate the concentration of an aqueous weak acid (or alkali) from its K_a (or K_b) and pH of the solution. This occurs with buffer solutions.

Here is an example of type 1.

Example: What is the pH of an aqueous solution of ethanoic acid of concentration $0.1\,mol\,dm^{-3}$? ($K_a = 1.7 \times 10^{-5}\,mol\,dm^{-3}$)

$$CH_3CO_2H(aq) + H_2O(l) \rightleftharpoons H_3O^+(aq) + CH_3CO_2^-(aq)$$
$$\frac{[H_3O^+(aq)][CH_3CO_2^-(aq)]}{[CH_3CO_2H(aq)]} = 1.7 \times 10^{-5}\,mol\,dm^{-3}$$

If the acid provides all the $H_3O^+(aq)$ ions then

$$[H_3O^+(aq)] = [CH_3CO_2^-(aq)]$$

If the acid dissociation into ions is negligible then

$$[CH_3CO_2H(aq)] = 0.1\,mol\,dm^{-3}$$
$$[H_3O^+(aq)]^2 = 0.1 \times 1.7 \times 10^{-5}\,mol^2\,dm^{-6}$$
$$[H_3O^+(aq)] = \sqrt{(1.7)} \times 10^{-3}\,mol\,dm^{-3}$$
$$pH = -\log_{10}([H_3O^+(aq)]/mol\,dm^{-3}) \quad pH = 2.9$$

Brønsted–Lowry theory

→ An acid is a molecule or ion that can donate a proton.
→ A base is a molecule or ion that can accept a proton.
→ A strong acid has a weak conjugate base and a weak acid has a strong conjugate base.

Acid–base strength

Acid	=	H^+	+	conjugate base		$K_a/mol\,dm^{-3}$		pK_a
$H_3O^+(aq)$				$H_2O(l)$		1.0		0
$H_2SO_3(aq)$				$HSO_3^-(aq)$		1.5×10^{-2}		1.8
$HSO_4^-(aq)$				$SO_4^{2-}(aq)$		1.02×10^{-2}		1.99
$HSO_3^-(aq)$				$SO_3^{2-}(aq)$		1.0×10^{-2}		2.0
$H_3PO_4(aq)$				$H_2PO_4^-(aq)$		7.9×10^{-3}		2.1
$HF(aq)$				$F^-(aq)$		5.6×10^{-4}		3.3
$HNO_2(aq)$				$NO_2^-(aq)$		4.7×10^{-4}		3.3
$CH_3CO_2H(aq)$				$CH_3CO_2^-(aq)$		1.7×10^{-5}		4.8
$H_2CO_3(aq)$				$HCO_3^-(aq)$		4.3×10^{-7}		6.4
$H_2PO_4^-(aq)$				$HPO_4^{2-}(aq)$		6.2×10^{-8}		7.2
$NH_4^+(aq)$				$NH_3(aq)$		5.6×10^{-10}		9.3
$C_6H_5OH(aq)$				$C_6H_5O^-(aq)$		1.3×10^{-10}		9.9
$HPO_4^{2-}(aq)$				$PO_4^{3-}(aq)$		4.4×10^{-13}		12.4
$H_2O(l)$				$OH^-(aq)$		1.0×10^{-14}		14.0

(Increasing Acid Strength ↑ / Increasing Base Strength ↓)

Exam practice   (5 minutes)   answers: page 68

Using the data above:

(a) What are the values of pK_b and K_b for ammonia?

(b) Explain why an aqueous solution of trisodiumphosphate, Na_3PO_4, would be extremely alkaline.

Acid–base reactions

An acid plus a base makes a salt plus water and we call this neutralization, but the salts we make are not always neutral! We find out why and then use this information to help us to choose the right indicator for a titration.

Watch out!

The reaction of an acid with metal is *not* a proton-transfer but an electron-transfer (redox) reaction.

Don't forget!

The lower the pK_a the stronger the acid.

Checkpoint 1

Explain how and why you can use sodium carbonate to distinguish between carboxylic acids and phenols.

Displacement reactions

➜ A strong acid with a low pK_a can transfer its protons to the conjugate base of a weaker acid with a higher pK_a.

Acids with $pK_a < 10.3$ are stronger than the hydrogencarbonate ion (acting as a proton donor) so they will react with its conjugate base (carbonate ion), e.g.

$$CH_3CO_2H(aq) + CO_3^{2-}(aq) \rightarrow CH_3CO_2^-(aq) + HCO_3^-(aq)$$

And acids with $pK_a < 6.4$ are stronger than carbonic acid so they will react with its conjugate base (hydrogencarbonate ion), e.g.

$$CH_3CO_2H(aq) + HCO_3^-(aq) \rightarrow CH_3CO_2^-(aq) + H_2CO_3(aq)$$

The displaced carbonic acid is unstable and readily decomposes into water and carbon dioxide.

➜ A strong base with a low pK_b can accept protons from the conjugate acid of a weaker base with a higher pK_b.

Alkali and alkaline earth metal hydroxides will release ammonia from ammonium salts, e.g.

$$NH_4^+(aq) + OH^-(aq) \rightarrow NH_3(aq) + H_2O(l)$$

Checkpoint 2

Explain why lumps of calcium oxide but not glass beads coated in concentrated sulphuric acid may be used to dry ammonia gas.

Hydrolysis of salts

A reaction of an acid with a base is called neutralization.

The jargon

When the pK_a is less than 7 we refer to *proton donors* and their conjugate bases as acids and their salts. When the pK_a is greater than 7 we refer to *proton acceptors* and their conjugate acids as bases and their salts.

Strong acids and strong bases

When hydrochloric acid reacts with aqueous sodium hydroxide according to the following equation, sodium chloride is formed and the pH of the solution is 7.

$$HCl(aq) + NaOH(aq) \rightarrow NaCl(aq) + H_2O(l)$$

But $HCl(aq) + H_2O(l) \rightarrow H_3O^+(aq) + Cl^-(aq)$ since the acid is strong and $NaOH(aq) \rightarrow Na^+(aq) + OH^-(aq)$ since the base is strong. If we omit the spectator ions $Na^+(aq)$ and $Cl^-(aq)$ the equation for the neutralization of any strong acid by any strong base may be written as

$$H_3O^+(aq) + OH^-(aq) \rightarrow H_2O(l)$$

Weak acids and weak bases

Checkpoint 3

Explain why we should not be surprised to find that an aqueous solution of ammonium ethanoate has a pH = 7.

➜ Aqueous salts of weak acids and strong bases have a pH > 7.
➜ Aqueous salts of strong acids and weak bases have a pH < 7.
➜ Aqueous salts with pH ≠ 7 are said to be hydrolyzed.

You should expect an aqueous solution of ammonium chloride to have a pH < 7 because the ammonium cation, $NH_4^+(aq)$, is a moderately good proton donor but the chloride anion, $Cl^-(aq)$, is a very weak base. You should also expect an aqueous solution of sodium ethanoate to have a pH > 7 because the sodium cation is not an acid and the ethanoate anion, $CH_3CO_2^-(aq)$, is a moderately good proton acceptor.

Acid–base titrations

Titrating is measuring with a burette the volume of acid (or alkali), of known concentration, needed to react exactly with a measured (by pipette) volume of another solution (in a conical flask) containing a small amount of an indicator.

Strong base into a strong acid

$0.1 \, mol \, dm^{-3} \, NaOH(aq)$ titrated into $20 \, cm^3 \, 0.1 \, mol \, dm^{-3} \, HCl(aq)$

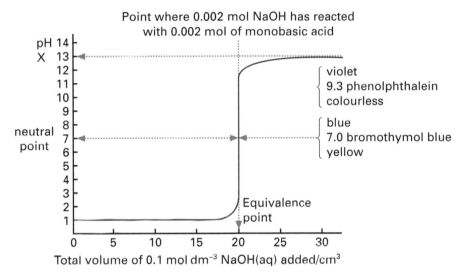

Strong base into a weak acid

$0.1 \, mol \, dm^{-3} \, NaOH(aq)$ titrated into $20 \, cm^3 \, 0.1 \, mol \, dm^{-3} \, CH_3CO_2H(aq)$

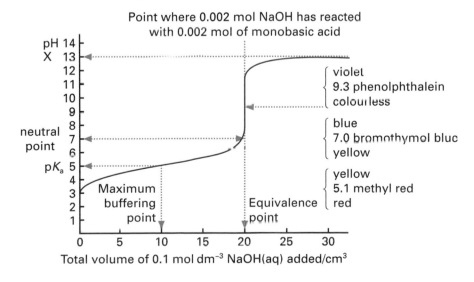

> **The jargon**
>
> *Basicity* is the number of moles of replaceable hydrogens in one mole of an acid. HCl is monobasic. H_2SO_4 is dibasic and H_3PO_4 is a tribasic acid.

> **Checkpoint 4**
>
> On these two diagrams what does the pH value at X correspond to?

> **The jargon**
>
> The *mid-point* colour of the indicator is where the pH = pK_{In}. The *end-point* is where the indicator shows its mid-point colour and the pH changes most for the smallest volume added from the burette. Ideally, the end-point and the equivalence point should be the same. Note that the end-point and equivalence point need not occur at the neutral point!

> **Checkpoint 5**
>
> Explain why the best indicator when titrating
> (a) strong acids with strong bases is bromothymol blue
> (b) weak acids by strong bases is phenolphthalein.

> **Grade booster**
>
> You need to be able to draw and interpret pH titration curves and indicators to gain the top grade.

Exam practice (5 minutes)   answers: page 68

Sketch a pH titration curve for titrating $20 \, cm^3 \, 0.1 \, mol \, dm^{-3} \, NH_3(aq)$ with $0.1 \, mol \, dm^{-3} \, HCl(aq)$ from a burette and explain why methyl red would be a suitable indicator.

Buffers and indicators

A buffer is a solution whose pH is almost unchanged by the addition of small amounts of acid or alkali. Buffers contain either a weak acid and its conjugate base (e.g. ethanoic acid and ethanoate ions) or a weak base and its conjugate acid (e.g. ammonia and ammonium ions).

Buffering action

If you add acid (H^+) to, say, an ethanoic acid/sodium ethanoate buffer, the conjugate base of the buffer combines with the added hydrogen ions: $H_3O^+(aq) + CH_3CO_2^-(aq) \rightarrow CH_3CO_2H(aq) + H_2O(l)$. If you add any alkali, the acid of the buffer combines with the added hydroxide ions: $CH_3CO_2H(aq) + OH^-(aq) \rightarrow CH_3CO_2^-(aq) + H_2O(l)$. For maximum buffering effect, the weak acid and conjugate base (or weak base and conjugate acid) should have the same concentrations, i.e. $[CH_3CO_2H(aq)] = [CH_3CO_2^-(aq)]$, so they cancel. Then

$$K_a = \frac{[H_3O^+(aq)][CH_3CO_2^-(aq)]}{[CH_3CO_2H(aq)]} = [H_3O^+(aq)]$$

and therefore

→ $pH = pK_a$ (or $pOH = pK_b$) at the maximum buffering point.

When a buffer is close to but not at its maximum buffering point

→ pH_{buffer} is governed by pK_a and the *ratio* of the concentrations of weak acid to conjugate base.
→ $pH_{buffer} = pK_a + \log_{10}([\text{conjugate base}]/[\text{weak acid}])$.

If you add only small amounts of dilute acid or alkali, the ratio [acid]/[conjugate base] changes very little and the logarithm of this ratio changes even less – so the buffer pH hardly changes at all. If you add large amounts of acid or alkali to a very dilute buffer solution the pH will change. A tiny drop of buffer won't cope with a bucket of concentrated acid!

→ The capacity of a buffer is governed by the *concentrations* of the weak acid and conjugate base.

Acid–base indicators

→ substances whose colour changes with pH
→ regarded as weak acids or weak bases
→ not very soluble so their coloured solutions are very dilute

Name of indicator	Weak acid form	pK_a	Conjugate base form
methyl orange	red	3.5	yellow
methyl red	red	5.1	yellow
bromothymol blue	yellow	7.0	blue
phenolphthalein	colourless	9.3	violet

Mid-point colour and pH range of an indicator

→ The dissociation constant of an indicator can be represented by

$$\frac{[H_3O^+(aq)][In^-(aq)]}{[HIn(aq)]} = K_{in}$$

→ The colour of an acid–base indicator is governed by the ratio of the concentrations of its weak acid and conjugate base.
→ An indicator shows its mid-point colour when [HIn(aq)] = [In⁻(aq)] and therefore when pH = pK_{in}.

If HIn(aq) is yellow and In⁻(aq) is blue, when [HIn(aq)] = [In⁻(aq)] the mid-point colour of the indicator solution will be green.

→ An acid–base indicator changes colour over a range of about 2 pH units, one unit either side of the pH at its mid-point colour.

Selecting an indicator

For an acid–base indicator to be suitable for a titration, pK_{in} should equal the pH at the end-point where pH changes most sharply.

0.1 mol dm⁻³ HCl(aq) titrated into 20 cm³ 0.1 mol dm⁻³ NH₃(aq)

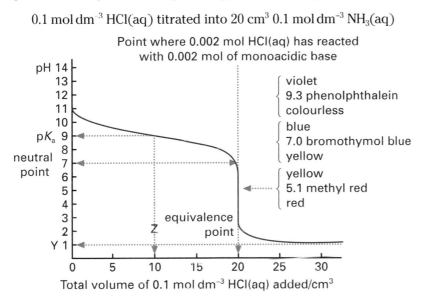

Point where 0.002 mol HCl(aq) has reacted with 0.002 mol of monoacidic base

We cannot titrate a weak acid with a weak alkali, or vice versa, because the pH does not change suddenly at the end-point so no indicator will give a sharp change in colour.

We avoid using concentrations less than 0.1 mol dm⁻³ because the range over which the pH will change rapidly gets shorter with decreasing concentrations of acid and base.

Exam practice   (7 minutes)   answers: page 69

Exam practice   (7 minutes)

(a) (i) Write an expression for the acid dissociation constant (K_a) for the equilibrium $NH_4^+(aq) \rightleftharpoons H^+(aq) + NH_3(aq)$ and (ii) calculate pK_a given that K_a is 5.6×10^{-10} mol dm⁻³ at 25 °C.

(b) Explain (i) why an aqueous solution containing both sodium dihydrogenphosphate, NaH_2PO_4, and disodium hydrogenphosphate, Na_2HPO_4, constitutes a buffer and (ii) why buffers are important in biological systems.

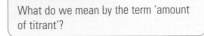

Checkpoint 2

What do we mean by the term 'amount of titrant'?

The jargon

NaOH and NH₃ are *monoacidic bases*. Ba(OH)₂ and NH₂CH₂CH₂NH₂ are *diacidic bases*.

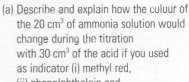

Checkpoint 3

(a) Describe and explain how the colour of the 20 cm³ of ammonia solution would change during the titration with 30 cm³ of the acid if you used as indicator (i) methyl red, (ii) phenolphthalein and (iii) bromothymol blue.
(b) On this diagram what does the pH value represent (i) at Y and (ii) at Z?

Action point

Sketch the variation in pH when
(i) aqueous sodium hydroxide of concentration 0.10 mol dm⁻³ is slowly added to 25 cm³ of hydrochloric acid of concentration 0.10 mol dm⁻³.
(ii) aqueous sodium hydroxide of concentration 0.10 mol dm⁻³ is slowly added to 25 cm³ of ethanoic acid of concentration 0.10 mol dm⁻³.
pK_a for ethanoic acid at laboratory temperature is 4.76.
(iii) Suggest a method by which the volume of a given aqueous solution of ammonia that reacts with a given volume of aqueous ethanoic acid could be determined practically.

Structured exam question

answer: page 69

(a) Many acid–base indicators are weak acids whose aqueous solution has a different colour from that of their conjugate base. The dissociation of the indicator molecules (HInd) into their conjugate bases (Ind⁻) may be represented by

$$HInd(aq) \rightleftharpoons H^+(aq) + Ind^-(aq)$$
$$\text{colour 1} \qquad \text{colour 2}$$

The pH range of such indicators is approximately $pK_{ind} \pm 1$.

(i) Show that $pH = pK_{ind} + \log([Ind^-(aq)]/[HInd(aq)])$

....

....

....

....

....

(ii) If HInd(aq) is yellow, Ind⁻(aq) is blue and pK_{ind} is 4.0, what is the pH of the indicator solution whose colour is green (the mid-point colour of the indicator)?

....

(iii) If HInd(aq) is yellow, Ind⁻(aq) is blue and pK_{ind} is 7.0, state and explain what colour would be formed on adding the indicator to 0.1 mol dm⁻³ NaOH(aq).

....

....

(b) (i) State what is meant by the term *buffer solution*.

....

....

....

(ii) Explain how an aqueous solution containing ethanoic acid and sodium ethanoate behaves as a buffer solution.

....

....

....

....

....

....

(iii) Calculate the pH of a buffer solution containing 0.10 mol dm⁻³ of ethanoic acid and 0.40 mol dm⁻³ of sodium ethanoate.
(The value of K_a for ethanoic acid is 1.8×10^{-5} mol dm⁻³)

....

....

....

....

(15 min)

Examiner's secrets

You might be given the Henderson–Hasselbalch equation in the form
$pH = pK_a + \log([base]/[acid])$ or
$pH = pK_a - \log([acid]/[base])$.
If not, just remember it is only an expression for Ka after taking logs of [HJ][base]/[acid] and rearranging.

Examiner's secrets

If the buffer had been aqueous ammonia and ammonium chloride then
$pH = pK_a - \log([NH_4^+(aq)]/[NH_3(aq)])$.

Answers
Equilibria

Checkpoints

1 (a) Melting point and freezing point are the same. Both terms refer to the temperature where the liquid and the solid phases are in equilibrium under the external pressure (normally atmospheric pressure).

(b) When ice is added to a drink, some of the ice melts and this takes energy which comes from the liquid, so the temperature of the liquid falls. As long as ice remains, any energy the liquid gains from the surroundings will be used up in melting the ice so that the temperature of the drink remains low.

(c) The molecules of the covalent compound naphthalene are held in the solid by weak van der Waals forces so the vapour pressure is high. The ions in sodium chloride are held in the lattice by strong electrostatic forces and so the vapour pressure is negligible.

2 (i) There would be no units for K_p or K_c.

(ii) K_c would have units $mol^{-2}\,dm^6$ and K_p would have units pressure $^{-2}$.

Exam practice

(a) The standard molar enthalpy change of vaporization is the heat change at constant pressure when one mole of liquid becomes one mole of vapour under standard thermodynamic conditions of 1 atm and 298 K.

(b) The polar water molecules are held in the liquid by van der Waals forces, permanent dipole–dipole attractions and hydrogen bonding but the non-polar hexane molecules are held in the liquid by van der Waals forces only. Therefore much more energy is needed to separate the water molecules and the boiling point is much higher.

(c) At a high altitude, the atmospheric pressure is much less than the normal 10.3 kPa at sea level. Therefore the vapour pressure of water reaches the value of the external pressure below 100 °C and the liquid boils.

Chemical equilibria

Checkpoints

1 High pressure favours a high equilibrium yield of ammonia since there are four moles of gas on the left-hand side of the equation and only two moles of gas on the right-hand side: $N_2(g) + 3H_2(g) \rightleftharpoons 2NH_3(g)$.

2 In the Contact process $2SO_2(g) + O_2(g) \rightleftharpoons 2SO_3(g)$, and at equilibrium $K_c = [SO_3]^2/[SO_2]^2[O_2]$. So more oxygen (excess) will mean less sulphur dioxide and more sulphur trioxide, i.e. more SO_2 converted into SO_3 and therefore a greater yield (equilibrium position is shifted to the right).

Exam practice

Le Chatelier's principle states that if a constraint (such as change in temperature, pressure or concentration) is applied to a system in equilibrium, then the equilibrium alters in such a way as to minimize the applied constraint.

$$K_p = \frac{(p_{PCl_3} \times p_{Cl_2})}{p_{PCl_5}}$$

(i) If the total pressure is increased, the equilibrium moves to the side of the equation with the fewer number of moles of gas. In this case, this means less dissociation and less PCl_3 and Cl_2 in the equilibrium mixture.

(ii) Since the reaction in the forward direction is a dissociation it must endothermic, ΔH is positive. Therefore, a rise in temperature will favour the endothermic process and more dissociation will take place.

Equilibrium problems

Checkpoint

Although high pressure favours the formation of two moles of $SO_3(g)$ from two moles of $SO_3(g)$ and one mole of $O_2(g)$, the equilibrium yield of sulphur trioxide is very high (about 95%) even at the minimum pressure needed to drive the gases through the plant. The small increase in yield would not justify the cost of high-pressure plant.

Exam practice

$CH_3COOH(l) + C_2H_5OH(l) \rightleftharpoons CH_3COOC_2H_5(l) + H_2O(l)$

initial   2 mol   1 mol

equil.   1.15 mol   0.15 mol   0.85 mol   0.85 mol

Therefore K_c is $\dfrac{[(0.85/V) \times (0.85/V)]}{[(1.15/V) \times (0.15/V)]} = 4.2$

Aqueous equilibria

Checkpoints

1 Since the ionization of water is endothermic, an increase in temperature will increase the extent of ionization.

2 pH is $-\log_{10}(2 \times 10^{-2}) = 1.7$

3 pOH $= -\log_{10}(0.02) = 1.7$

pH $= pK_w - pOH$

pH $= 14 - 1.7 = 12.3$

4 The gas is carbon dioxide.
Limewater is saturated aqueous calcium hydroxide. The acidic gas reacts with hydroxide ions to form aqueous carbonate ions that combine with the aqueous calcium ions to give the milky white precipitate of insoluble calcium carbonate:

$CO_2(g) + 2OH^-(aq) + Ca^{2+}(aq) \rightarrow CaCO_3(s) + H_2O(l)$

Exam practice

(a) $pH = -\log_{10}([H^+(aq)]/mol\,dm^{-3})$
or $pH = -\log_{10}([H_3O^+(aq)]/mol\,dm^{-3})$

(b) You could give at least two answers. You could explain that sulphuric acid is a strong monobasic acid, so
$$H_2SO_4(aq) \rightarrow H^+(aq) + HSO_4^-(aq)$$
Then $[H^+(aq)] = 0.05\,mol\,dm^{-3}$ and pH is
$-\log_{10}([H^+(aq)]/mol\,dm^{-3}) = 1.3$.
Or you could explain that sulphuric acid is a dibasic acid,
$$H_2SO_4(aq) \rightarrow 2H^+(aq) + SO_4^{2-}(aq)$$
Then $[H^+(aq)] = 2 \times 0.05\,mol\,dm^{-3}$ and pH is
$-\log_{10}([H^+(aq)]/mol\,dm^{-3}) = 1.0$.
You would get full marks for either answer.
If you also explained that there is not enough information to solve this problem accurately because the second ionization $HSO_4^-(aq) \rightleftharpoons H^+(aq) + SO_4^{2-}(aq)$ would occur to some extent, you might get a bonus mark.

(c) If pH = 10.8 then pOH is 14 − 10.8 = 3.2. So,
$-\log_{10}([OH^-(aq)]/mol\,dm^{-3}) = 3.2$
$[OH^-(aq)] = 6.3 \times 10^{-4}\,mol\,dm^{-3}$
One mole of $Ca(OH)_2$ produces two moles of OH^-.
So the concentration of $Ca(OH)_2 = 3.15 \times 10^{-4}\,mol\,dm^{-3}$.
$M_r(Ca(OH)_2)$ is 40 + (2 × 17) = 74
So the concentration of $Ca(OH)_2$ is
$3.15 \times 10^{-4} \times 74 = 2.3 \times 10^{-2}\,g\,dm^{-3}$.

Weak acids and bases

Exam practice

(a) pK_a for ammonia is 9.3
Therefore pK_b is 14 − 9.3 = 4.7
pK_b is $-\log K_b = 4.7$
So, K_b is $2.0 \times 10^{-5}\,mol\,dm^{-3}$ (*not* 1.99×10^{-5}).

(b) When the ionic compound $(Na^+)_3PO_4^{3-}$ dissolves in water, it releases PO_4^{3-}, a very strong base which disturbs the $H_2O(l) \rightleftharpoons H^+(aq) + OH^-(aq)$ equilibrium by removing $H^+(aq)$ to form HPO_4^{2-}. $[H^+(aq)] < [OH^-(aq)]$, pH is greater than 7 and the solution is alkaline.

Acid–base reactions

Checkpoints

1 Aqueous carboxylic acids are stronger acids than aqueous carbon dioxide; therefore when a carboxylic acid is added to aqueous sodium carbonate, carbon dioxide is liberated.
Phenols in aqueous solution are weaker acids than aqueous carbon dioxide and so no reaction takes place.

2 Ammonia is a basic gas and would not react with another base like calcium oxide (which removes moisture from the ammonia by forming calcium hydroxide) but it would react with sulphuric acid.
$$NH_3 + H_2SO_4 \rightarrow NH_4HSO_4$$

3 pK_a for ethanoic acid is about 5 and pK_b for aqueous ammonia is about 5. So although the acid and base are both weak, they are weak to a similar extent. Therefore an aqueous solution of their salt is likely to be nearly neutral, pH = 7.

4 The pH value of the aqueous sodium hydroxide.

5 (a) Bromothymol blue will change colour as soon as the pH changes to a value greater than 7, i.e. it will change colour at the equivalence point.

(b) Phenolphthalein changes colour in the pH range covered by the steep portion of the curve. This means that one drop of alkali from the burette will cause the pH value to change by several pH units and the colour of the indicator to change sharply from colourless to pink.

Exam practice

pH range curve:

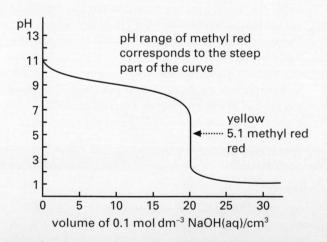

Buffers and indicators

Checkpoints

1 No effect since the ratio [salt]/[acid] is constant.

2 The volume of liquid added from the burette.

3 (a) (i) Initially the ammonia solution containing methyl red would be yellow and remain yellow until near the equivalence point when it would become orange, until one drop of acid turned the solution red. No further change in colour thereafter.

(ii) The ammonia solution would remain red until about 18 cm³ of acid had been added when it would change to colourless. After that no further colour change.

(iii) The ammonia solution containing bromothymol blue would start blue and when almost 20 cm³ of acid had been added it would turn yellow and then stay yellow.

(b) Y is the pH value of the aqueous hydrochloric acid. Z is the volume of acid which gives a pH value equal to pK_a ($= 9$) for the *ammonium ion*. (NB pK_b is $14 - 9 = 5$ and $K_b = 1.0 \times 10^{-5}$ mol dm⁻³ for aqueous ammonia.)

Exam practice

(a) (i) $K_a = \dfrac{[H^+][NH_3]}{[NH_4^+]}$

(ii) $pK_a = -\log_{10}(5.6 \times 10^{-10}) = 9.25$

(b) (i) The hydrogenphosphate ion HPO_4^{2-}(aq) is the conjugate base of the dihydrogenphosphate ion $H_2PO_4^-$(aq). So a mixture of the two salts will work as a buffer just as ethanoic acid and its conjugate base, the ethanoate ion, act as a buffer. HPO_4^{2-}(aq) will accept protons from any acid added and $H_2PO_4^-$(aq) will donate protons to any base added to the buffer.

(ii) Buffers are important in biological systems since many biochemical processes are pH sensitive. Some enzymes only work within a narrow pH range. The blood is buffered within the range pH 7.39 – 7.41 by carbon dioxide, water and hydrogencarbonate ions.

Structured exam question

(a) (i) Applying the law of chemical equilibrium to the reversible reaction $HInd(aq) \rightleftharpoons H^+(aq) + Ind^-(aq)$ gives

$$\frac{[H^+(aq)][Ind^-(aq)]}{[HInd(aq)]} = K_{ind}$$

Taking logarithms:
$\log[H^+(aq)] + \log([Ind^-(aq)]/[HInd(aq)]) = \log K_{ind}$

Hence:
$-pH + \log([Ind^-(aq)]/[HInd(aq)]) = -pK_{ind}$

Rearranging:
$-pH = -pK_{ind} - \log([Ind^-(aq)]/[HInd(aq)])$

Hence:
$pH = pK_{ind} + \log([Ind^-(aq)]/[HInd(aq)])$

(ii) Equal intensities of yellow and blue give green mid-point colour which means that $[HInd(aq)] = [Ind^-(aq)]$ when pH is 4.0. So $\log([Ind^-(aq)])/[HInd(aq)])$ is $\log 1$ ($= 0$) and $pH = pK_{ind} = 4.0$.

(iii) Blue. In 0.1 mol dm⁻³ NaOH(aq) the pH $= 13$ (because pOH $= 1$ and pH is $14 - pOH$), and so
$13 = 7 + \log([Ind^-(aq)]/[HInd(aq)])$
If $\log([Ind^-(aq)]/[HInd(aq)]) = 6$, then
$[Ind^-(aq)] = 10^6 \times [HInd(aq)]$.
Blue outweighs yellow by 1 million to 1.

(b)(i) A solution of a weak acid and its conjugate base (or weak base and its conjugate acid) whose pH is almost unchanged by the addition of small amounts of acid or alkali.

(ii) In the buffer solution
$CH_3COOH(aq) \rightleftharpoons CH_3COO^-(aq) + H^+(aq)$
so ethanoic acid molecules are in equilibrium with ethanoate ions. When acid is added, momentarily increasing the $[H^+(aq)]$, the conjugate base reacts with the added $H^+(aq)$ and the equilibrium shifts to the left.
When alkali is added, momentarily increasing the $[OH^-(aq)]$, the acid reacts with the added $OH^-(aq)$ and the equilibrium shifts to the right.

(iii) $pH = pK_a + \log([CH_3COO^-(aq)]/[CH_3COOH(aq)])$
$= -\log(1.8 \times 10^{-5}) + \log(0.40/0.10)$
Thus pH $= 5.3$.

Electrochemistry and redox

You are already familiar with oxidation numbers in chemical names like copper(II) sulphate. In this chapter you will learn the simple rules for assigning these numbers. You will also understand the fundamentals of electrochemical cells and the importance of electrode potentials. In the previous section you saw acid–base reactions as the transfer of protons from acids to bases. In this section you see redox reactions as the transfer of electrons from reducing agents to oxidizing agents.

By the end of this chapter you will be able to

- Work out oxidation numbers and use them to identify redox reactions and balance redox equations

- Describe how to measure the EMF of a cell

- Describe the hydrogen electrode and define what is meant by standard thermodynamic conditions

- Draw and interpret cell diagrams

Topic checklist

Tick each of the boxes below when you are satisfied that you have mastered the topic

	Edexcel		AQA		OCR		WJEC		CCEA	
	AS	A2	AS	A2	AS	A2	AS	A2	AS	A2
Oxidation and reduction	○	●	○	●	○	●	○	●	○	●
Electrochemistry		●		●		●		●		●
Electrochemical series		●		●		●		●		●

Oxidation and reduction

Oxidation can be spotted in three different ways: gain of oxygen, loss of electrons or an increase in oxidation number. Not surprisingly, reduction is the reverse. We cannot have reduction without oxidation and we call reactions involving reduction and oxidation redox reactions.

Oxidation numbers

The oxidation number is a number assigned to an element according to this set of rules applied in the following priority order:

1	oxidation number of an uncombined element	0
2	sum of oxidation numbers of elements in uncharged formula	0
3	sum of oxidation numbers of elements in charged formula	charge
4	oxidation number of fluorine in any formula	−1
5	oxidation number of an alkali metal in any formula	+1
6	oxidation number of an alkaline earth metal in any formula	+2
7	oxidation number of oxygen (except in peroxides = −1)	−2
8	oxidation number of halogen in metal halides	−1
9	oxidation number of hydrogen (except in metal hydrides = −1)	+1

Redox reactions

→ Reduction is the gain of electrons by an oxidant: e.g.

$$Cl_2(aq) + 2e^- \rightarrow 2Cl^-(aq)$$

→ Oxidation is the loss of electrons by a reductant, e.g.

$$Fe^{2+}(aq) \rightarrow Fe^{3+}(aq) + e^-$$

→ Redox is the transfer of electrons from a reductant to an oxidant:

$$2Fe^{2+}(aq) + Cl_2(aq) \rightarrow 2Fe^{3+}(aq) + 2Cl^-(aq)$$

A redox reaction involves an increase (↑) in oxidation number of one element (in the reductant) and a simultaneous balancing decrease (↓) in oxidation number of an element (in the oxidant).

→ Disproportionation is a redox involving a simultaneous increase and decrease in oxidation number of the same element:

$$Cl_2(aq) + 2OH^-(aq) \rightarrow ClO^-(aq) + Cl^-(aq) + H_2O(l)$$

This disproportionation equation may also be written in the form

$$Cl_2(aq) + H_2O(l) \rightarrow ClO^-(aq) + Cl^-(aq) + 2H^+(aq)$$

Writing $H_2O(l)$ on the LHS and $2H^+(aq)$ on the RHS has the same effect as wvriting $2OH^-(aq)$ on the LHS and $H_2O(l)$ on the RHS because hydrogen ions react with hydroxide ions to form water molecules and the reaction $H^+(aq) + OH^-(aq) \rightarrow H_2O(l)$ is acid–base (*not* redox).

Balancing redox reactions

You can use the changes in oxidation numbers to work out balanced equations for redox reactions.

Example: Reaction of acidified dichromate(VI) with sulphate(IV).

1 Write down correct formulae of reactants and products:
$$Cr_2O_7^{2-}(aq) + H^+(aq) + SO_3^{2-}(aq) \rightarrow 2Cr^{3+}(aq) + SO_4^{2-}(aq)$$

2 Apply priority rules to assign oxidation numbers:
$$Cr_2O_7^{2-}(aq) + H^+(aq) + SO_3^{2-}(aq) \rightarrow 2Cr^{3+}(aq) + SO_4^{2-}(aq)$$
$$+6 \qquad\qquad +4 \qquad +3 \qquad +6$$

3 Calculate rise and fall of oxidation numbers:
$$Cr_2O_7^{2-}(aq) + H^+(aq) + SO_3^{2-}(aq) \rightarrow 2Cr^{3+}(aq) + SO_4^{2-}(aq)$$
rise of 2

answer: page 79

4 Balance the rise and fall of oxidation numbers:
$$Cr_2O_7^{2-}(aq) + H^+(aq) + 3SO_3^{2-}(aq) \rightarrow 2Cr^{3+}(aq) + 3SO_4^{2-}(aq)$$
$$3 \times 2 = 6$$

Don't forget!

If you want the top grade make sure you balance the rise and fall of the oxidation numbers because the electrons lost by the reductant are exactly the same as the electrons gained by the oxidant.

5 Balance other atoms *without changing balance of redox atoms*:
$$Cr_2O_7^{2-}(aq) + H^+(aq) + 3SO_3^{2-}(aq) \rightarrow 2Cr^{3+}(aq) + 3SO_4^{2-}(aq)$$
7 oxygens 9 oxygens 12 oxygens

(i) Write $4H_2O(l)$ on the RHS to provide 4 more oxygens.
(ii) Write $8 H^+(aq)$ on the LHS to balance the hydrogens.

6 Check the charges balance (same net charge on LHS and RHS):
$$LHS(2-) + 8 \times (1+) + 3 \times (2-) = 0$$
$$Cr_2O_7^{2-}(aq) + 8H^+(aq) + 3SO_3^{2-}(aq)$$
$$\rightarrow 2Cr^{3+}(aq) + 3SO_4^{2-}(aq) + 4H_2O(l)$$
$$RHS\ 2 \times (3+) + 3 \times (2-) = 0$$

Redox titrations and calculations

The manganate(VII) ion in *excess* dilute sulphuric acid is *quantitatively* reduced to the manganese(II) ion by $Fe^{2+}(aq)$, $H_2O_2(aq)$, $C_2O_4^{2-}(aq)$, $SO_3^{2-}(aq)$, etc., and the solution changes from purple to colourless.

Consequently, we may titrate $MnO_4^-(aq)$ of known concentration from a burette into a *measured volume* of these reducing agents *without adding an indicator* to determine their concentrations.

Grade booster

You should do titrations with potassium manganate(VII) and sodium thiosulphate solutions and be able to use your results in calculations.

Checkpoint 2

Explain, in manganate(VII) titrations,
(a) why we use excess acid
(b) what we mean by quantitatively
(c) how we get measured volumes of the reducing agent, and
(d) why we do not need to add an indicator.

Example: 20.0 cm^3 of aqueous iron(II) sulphate need 18.5 cm^3 of 0.105 mol dm^{-3} KMnO$_4$(aq). What is the concentration of the iron(II) sulphate?

$$MnO_4^-(aq) + 8H^+(aq) + 5Fe^{2+}(aq) \rightarrow Mn^{2+}(aq) + 4H_2O(l) + 5Fe^{3+}(aq)$$
1 mol 5 mol
$18.5 \times 0.1/1\ 000$ $5 \times (18.5 \times 0.105/1\ 000)$ mol in 20.0 cm^3 solution

So in 1 000 cm^3 solution $5 \times (18.5 \times 0.105/20.0) = 0.486$ mol Fe^{2+}(aq).
1 mol iron(II) sulphate, FeSO$_4$, contains 1 mol iron(II) ions, Fe^{2+}.
Concentration of the iron(II) sulphate is 0.486 mol dm^{-3}.

Watch out!

Do a quick mental check on your answer – round 18.5 to 20 – then the equation tells you the iron(II) concentration should be about 5 × the manganate(VII) concentration. But you rounded 18.5 *up* to 20 so the iron(II) concentration comes *down* below 5 × 0.105.

Exam practice (3 min) answer: page 79

Explain the type of reaction represented by the following equations:

(a) $Mg(s) + 2HCl(aq) \rightarrow MgCl_2(aq) + H_2(g)$
(b) $2K_2CrO_4(aq) + H_2SO_4(aq) \rightarrow K_2Cr_2O_7(aq) + K_2SO_4(aq) + H_2O(l)$

Electrochemistry

We can write separate ion–electron half-reaction equations for reduction and oxidation. We can also combine these half-equations to obtain an ionic equation for the redox.

Metal displacement reactions

Coating a nail by dipping it into aqueous copper(II) sulphate is a simple example of an electrochemical (redox) reaction.

$$Cu^{2+}(aq) + 2e^- \rightarrow Cu(s) \qquad \text{reduction}$$
$$Fe(s) \rightarrow Fe^{2+}(aq) + 2e^- \qquad \text{oxidation}$$
$$Fe(s) + Cu^{2+}(aq) \rightarrow Fe^{2+}(aq) + Cu(s) \qquad \text{redox}$$

Coating a piece of zinc by dipping it into aqueous iron(II) sulphate is another example of a metal displacement (redox) reaction.

$$Fe^{2+}(aq) + 2e^- \rightarrow Fe(s) \qquad \text{reduction}$$
$$Zn(s) \rightarrow Zn^{2+}(aq) + 2e^- \qquad \text{oxidation}$$
$$Zn(s) + Fe^{2+}(aq) \rightarrow Zn^{2+}(aq) + Fe(s) \qquad \text{redox}$$

We can use these reactions to produce an electric current.

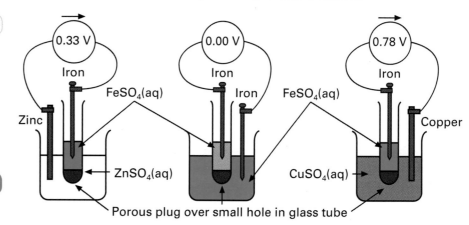

We can also use the displacement reaction of zinc with copper(II) ions.

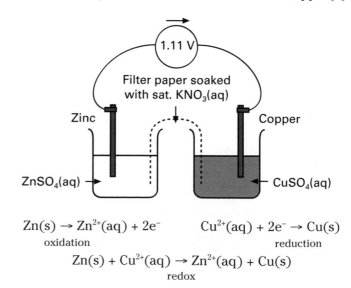

$$Zn(s) \rightarrow Zn^{2+}(aq) + 2e^- \qquad Cu^{2+}(aq) + 2e^- \rightarrow Cu(s)$$
$$\text{oxidation} \qquad\qquad\qquad\qquad \text{reduction}$$
$$Zn(s) + Cu^{2+}(aq) \rightarrow Zn^{2+}(aq) + Cu(s)$$
$$\text{redox}$$

Zinc and iron (but not copper) will also react with hydrogen ions to form hydrogen gas, e.g. $Zn(s) + 2H^+(aq) \rightarrow Zn^{2+}(aq) + H_2(g)$.

Reactivity series of metals

On the basis of their displacement reactions we can arrange the metals (and hydrogen) in reactivity order and their cations in stability order.

from most reactive Zn(s)   Fe(s)   H$_2$(g)   Cu(s)   to least reactive
from most stable   Zn^{2+}(aq)   Fe^{2+}(aq)   H$_3$O$^+$(aq)   Cu^{2+}(aq)   to least stable

Electromotive force of electrochemical cells

→ The e.m.f. (E) of an electrochemical cell is the maximum potential difference (voltage) between the electrodes.

→ The *standard* e.m.f. ($E^{\ominus}$) is the maximum voltage of a cell under standard conditions, i.e. temperature 25 °C (298 K), pressure 1 atm (101 kPa) and solutions of concentration 1 mol dm^{-3}.

Apparatus diagrams and cell diagrams

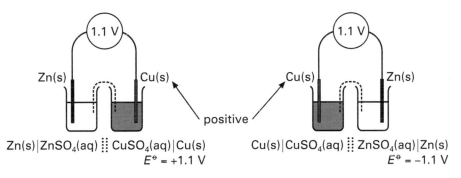

Zn(s)|ZnSO$_4$(aq) ⁞ CuSO$_4$(aq)|Cu(s)
$E^{\ominus}$ = +1.1 V

Cu(s)|CuSO$_4$(aq) ⁞ ZnSO$_4$(aq)|Zn(s)
$E^{\ominus}$ = −1.1 V

The sign given to the e.m.f. associated with the cell diagram of an electrochemical cell refers to the right-hand half-cell in the cell diagram.

Standard hydrogen electrode

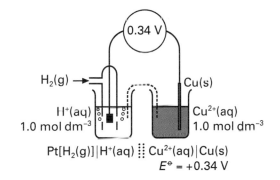

Pt[H$_2$(g)]|H$^+$(aq) ⁞ Cu^{2+}(aq)|Cu(s)
$E^{\ominus}$ = +0.34 V

The standard hydrogen electrode is the half-cell consisting of pure hydrogen gas, at 25 °C and 1 atm pressure, bubbling past a platinized platinum electrode dipping into 1.0 mol dm^{-3} H$_3$O$^+$(aq).

The above value of +0.34 V is called the standard electrode potential of the system Cu^{2+}(aq)|Cu(s).

Exam practice (2 min)   answer: page 79

(a) For the cell shown above write the ion–electron half-equation for
    (i) the oxidation of the hydrogen gas and (ii) the reduction of the
    aqueous copper(II) ions.

(b) Combine the half-equations in (a)(i) and (ii) to write the ionic equation
    for cell redox.

Don't forget!

The maximum voltage (e.m.f.) corresponds to zero current through the cell. It must be measured with a high resistance (valve) voltmeter or a potentiometer.

Watch out!

Do not confuse the diagram of a cell (drawing of the apparatus) with a cell diagram (symbolic representation of the electrochemistry).

The jargon

In a cell diagram, a vertical line (|) represents a (phase) boundary between a solid and a solution. A pair of vertical broken lines represents a salt bridge electrolytic boundary between two solutions.

The jargon

Platinized platinum: finely divided platinum is deposited electrolytically as a black coating on the platinum surface to improve contact between the metal, the gas and the solution and to catalyse the oxidation half-reaction.

Electrochemical series

We can combine two half-cells to form a complete electrochemical cell. If one half-cell is a hydrogen electrode then the e.m.f. of the complete cell is called the electrode potential of the other half-cell.

The jargon

Equations such as
$Cl_2(aq) + 2e^- \rightarrow 2Cl^-(aq)$
are called ion/electron half-equations.

Checkpoint

(a) What is the colour of (i) aqueous potassium halides (ii) KBr(aq) and KI(aq) after reaction with chlorine?
(b) Write the ordinary equation for the reaction of aqueous bromine with aqueous potassium iodide.

The jargon

The graphite rods (like platinized platinum) are called *inert electrodes* because (unlike copper and zinc) they are not part of the electrochemical reaction.

Watch out!

We have written $F_2(g)$ *not* $F_2(aq)$ because fluorine is a more powerful oxidizer than oxygen. So it attacks water to form hydrofluoric acid and displaces the oxygen.

Non-metal displacement reactions

Adding chlorine to aqueous potassium bromide or iodide produces bromine or iodine and changes the colour of the solution.

$$Cl_2(aq) + 2e^- \rightarrow 2Cl^-(aq) \qquad \text{reduction}$$
$$2Br^-(aq) \rightarrow Br_2(aq) + 2e^- \qquad \text{oxidation}$$
$$2Br^-(aq) + Cl_2(aq) \rightarrow Br_2(aq) + 2Cl^-(aq) \qquad \text{redox}$$

We can use these displacement reactions to produce an electric current.

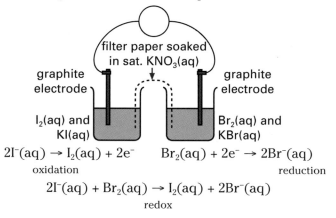

filter paper soaked in sat. $KNO_3(aq)$

graphite electrode   graphite electrode

$I_2(aq)$ and KI(aq) $Br_2(aq)$ and KBr(aq)

$2I^-(aq) \rightarrow I_2(aq) + 2e^-$ $Br_2(aq) + 2e^- \rightarrow 2Br^-(aq)$
oxidation   reduction

$$2I^-(aq) + Br_2(aq) \rightarrow I_2(aq) + 2Br^-(aq)$$
redox

Reactivity series of non-metals

On the basis of their displacement reactions we can arrange the halogens in reactivity order and their anions in stability order.

from most reactive	$F_2(g)$	$Cl_2(aq)$	$Br_2(aq)$	$I_2(aq)$	to least reactive
from most stable	$F^-(aq)$	$Cl^-(aq)$	$Br^-(aq)$	$I^-(aq)$	to least stable

If we want to judge the reactivity of metals (reductants) and non-metals (oxidants) on the same basis, we should rewrite the reactivity table for non-metal anions acting as reductants

from most reactive	$I^-(aq)$	$Br^-(aq)$	$Cl^-(aq)$	$F^-(aq)$	to least reactive
from most stable	$I_2(aq)$	$Br_2(aq)$	$Cl_2(aq)$	$F_2(aq)$	to least stable

Reactivity series of metals and non-metals

If we judge the reactivity of metals and aqueous non-metal anions as reductants losing electrons, we may combine the two series into one.

from most reactive reductant							to least reactive reductant
Zn(s)	Fe(s)	$H_2(g)$	Cu(s)	$I^-(aq)$	$Br^-(aq)$	$Cl^-(aq)$	$F^-(aq)$
$Zn^{2+}(aq)$	$Fe^{2+}(aq)$	$H_3O^+(aq)$	$Cu^{2+}(aq)$	$I_2(aq)$	$Br_2(aq)$	$Cl_2(aq)$	$F_2(aq)$
from least reactive oxidant							to most reactive oxidant

From this series we predict that zinc and fluorine would react best to form the most stable product, zinc fluoride. In general we predict that any reductant on the left could react with any oxidant to its right.

Standard electrode potentials

The **standard electrode potential**, $E^\ominus$, is defined as the e.m.f. of an electrochemical cell in which a standard hydrogen electrode is shown as the left-hand half-cell.

The standard electrode potential for the standard hydrogen electrode is zero by definition. A typical cell diagram for the standard electrode potential of a system is $Pt[H_2(g)]|2H^+(aq) \vdots Zn^{2+}(aq)\,Zn(s)$; $E^{\ominus} = -0.76$ V. Data books usually give the standard reduction potential (in volts) alongside the electrode system for the right-hand half-cell.

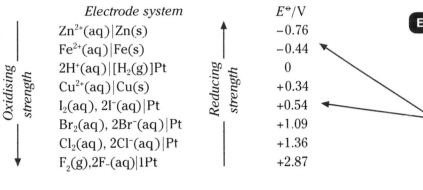

Electrode system	$E^{\ominus}$/V
$Zn^{2+}(aq)|Zn(s)$	−0.76
$Fe^{2+}(aq)|Fe(s)$	−0.44
$2H^+(aq)|[H_2(g)]Pt$	0
$Cu^{2+}(aq)|Cu(s)$	+0.34
$I_2(aq), 2I^-(aq)|Pt$	+0.54
$Br_2(aq), 2Br^-(aq)|Pt$	+1.09
$Cl_2(aq), 2Cl^-(aq)|Pt$	+1.36
$F_2(g), 2F{-}(aq)|1Pt$	+2.87

(left axis: Oxidising strength ↓; right axis: Reducing strength ↑)

Combining electrode potentials

You should be able to use tables of standard electrode potentials to predict feasible reactions, write cell diagrams and calculate standard e.m.f. values. Here is a simple example using the table above.

RHS: reductant $Zn(s)$ is above (more powerful than) $2I^-(aq)$
LHS: oxidant $I_2(aq)$ is below (more powerful than) $Zn^{2+}(aq)$

Therefore predict $Zn(s) + I_2(aq) \rightarrow Zn^{2+}(aq) + 2I^-(aq)$.

cell diagram $Zn(s)|Zn^{2+}(aq) \vdots I_2(aq), 2I^-(aq)|Pt$
e.m.f. of cell is $(+0.54 \text{ V}) - (-0.76 \text{ V}) = +1.30$ V

Spontaneous redox reactions

→ If the e.m.f. is not zero then the electrochemical reaction will take place in the cell when its terminals are wired together.
→ Electrons will travel in the wire from the negative half-cell losing them to the positive half-cell gaining them.

Example

$$2e^-$$
$$Zn(s)|Zn^{2+}(aq) \vdots Pb^{2+}(aq)|Pb(s) \quad E^{\ominus}_{cell} = +0.63 \text{ V}$$

Spontaneous cell reaction: $Zn(s) + Pb^{2+}(aq) \rightarrow Zn^{2+}(aq) + Pb(s)$

→ If a cell diagram corresponds to a positive e.m.f. the spontaneous cell reaction can be read from left to right.

$$Pb(s)|Pb^{2+}(aq) \vdots Zn^{2+}(aq)|Zn(s) \quad E^{\ominus}_{cell} = -0.63 \text{ V}$$

If a cell diagram corresponds to a negative e.m.f. the spontaneous cell reaction can be read from right to left.

Structured exam question

answers: pages 79–80

(a) Define the term *standard electrode potential* as applied to a metallic element.

....

....

(b) The standard electrode potentials $E^{\ominus}$ of zinc and copper are –0.76 V and +0.34 V respectively.

(i) Draw an apparatus diagram to show a cell in which zinc is in contact with aqueous zinc sulphate in a beaker and copper in contact with aqueous copper(II) sulphate in a similar beaker.

Watch out!

2 marks = 2 minutes for a neat, clearly labelled drawing. (b) (ii) may only take a few seconds, so you might be tempted to spend the time you save on your diagram. *Not* a good plan.

(ii) State the e.m.f. of this cell:..volts.

(iii) Name the instrument that would be used to measure this e.m.f.

....

(iv) In which direction would electrons flow in a wire used to connect together the copper and zinc electrodes?

....

(v) Write an equation for the overall spontaneous cell reaction that would occur when the copper and zinc electrodes are connected together by a wire.

....

(c) The table below shows standard redox potentials of selected reductions.

Ion–electron half-equation	$E^{\ominus}$/V
$BrO_3^- + 6H^+ + 5e^- \rightleftharpoons \frac{1}{2}Br_2 + 3H_2O$	+1.52
$\frac{1}{2}I_2 + e^- \rightleftharpoons I^-$	+0.54
$Fe^{3+} + e^- \rightleftharpoons Fe^{2+}$	+0.77
$Ag^+ + e^- \rightleftharpoons Ag$	+0.80

Predict, giving reasons, what reaction (if any) could occur in *each* of the following:

(i) Aqueous potassium iodide is added to acidified potassium bromate(V).

....

....

(ii) Metallic silver is added to aqueous iron(III) sulphate.

....

....

(iii) Aqueous iron(III) chloride and aqueous potassium iodide are mixed.

....

....

....

....

....

....

(15 min)

Answers
Electrochemistry and redox

Oxidation and reduction

Checkpoints

1 (a)(i) iron(II) ion
   (ii) iron(III) chloride
   (iii)chlorate(I) ion
  (b)(i) +5
   (ii) +7
  (c) $2Na(s) + 2H_2O(l) \rightarrow 2NaOH(aq) + H_2(g)$

It is redox because a sodium atom loses an electron to become a sodium ion. The oxidation number of sodium increases from 0 to +1. The oxidation number of hydrogen decreases from +1 to 0.

2 (a) Excess acid is used to ensure that all the manganate(VII) is reduced to manganese(II) ions. If insufficient acid is used some manganate(VII) ions are only reduced as far as the +4 state.

  (b) 'Quantitatively' means in exact measured amounts (moles) of substances according to the equation.
  (c) With a pipette, which delivers an exact volume of liquid (the solution of reductant in this case).

  (d) No indicator is needed because a solution containing aqueous manganate(VII) ions is purple (pink when very dilute) but turns colourless when reduced to a solution containing aqueous manganese(II) ions.

Exam practice

(a) This is a redox reaction.
   Reduction
   $$2H^+(aq) + 2e^- \rightarrow H_2(g)$$
   Oxidation
   $$Mg(s) \rightarrow Mg^{2+}(aq) + 2e^-$$
(b) This is an acid–base (proton-transfer) reaction:
   $$2CrO_4^{2-}(aq) + 2H^+(aq) \rightleftharpoons Cr_2O_7^{2-}(aq) + H_2O(l)$$
   It is *not* a redox reaction because the oxidation state of chromium (+6) does not change. The chromate(VI) ion is a Brønsted–Lowry base.

Electrochemistry

Checkpoint

(a) So that there is electrical contact (by the ions present) between the two solutions.
(b) The porous plug prevents rapid mixing of the two solutions.

Exam practice

(a)(i) $H_2(g) \rightarrow 2H^+(aq) + 2e^-$
   or $H_2(g) + 2H_2O(l) \rightarrow 2H_3O^+(aq) + 2e^-$
   (ii)$Cu^{2+}(aq) + 2e^- \rightarrow Cu(s)$
(b)$H_2(g) + Cu^{2+}(aq) \rightarrow 2H^+(aq) + Cu(s)$
   or $H_2(g) + 2H_2O(l) + Cu^{2+}(aq) \rightarrow 2H_3O^+(aq) + Cu(s)$

Electrochemical series

Checkpoint

(a)(i) Colourless.
   (ii)Aqueous potassium bromide goes brown due to liberated bromine. Aqueous potassium iodide first goes brown due to the formation of iodine which reacts with iodide ions to give the brown $I_3^-(aq)$ ion. If the chlorine is in excess then a black precipitate of iodine is seen.
(b)$Br_2(aq) + 2KI(aq) \rightarrow 2KBr(aq) + I_2(aq)$

Exam practice

(a)$Br_2(aq) + H_2(g) \rightarrow 2Br^-(g) + 2H^+(aq)$
   Br^2 is a stronger oxidant than H^+ and H_2 is a stronger reducer than Br^-
(b)no reaction
   Fe and I^- are both reductants
(c)no reaction
   H_3O^+ and Cl_2 are both oxidants
(d)no reaction
   Cl_2 and Cu^{2+} are both oxidants
(e)$Cu(s) + Br_2(aq) \rightarrow Cu^{2+}(aq) + 2Br^-(aq)$
   Br_2 is an oxidant and Cu is a reductant

Structured exam question

(a)The e.m.f. of an electrochemical cell under standard conditions of temperature (298 K), pressure (1 atm) and electrolyte concentration (1 mol dm^{-3}) represented by the cell diagram

$$Pt[H_2(g)] \mid 2H^+(aq) \ ⫶ \ M^{n+}(aq) \mid M(s)$$

where M(s) and $M^{n+}(aq)$ represent the metal and its cations.

(b) (i)

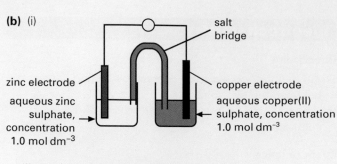

zinc electrode

aqueous zinc sulphate, concentration 1.0 mol dm^{-3}

salt bridge

copper electrode

aqueous copper(II) sulphate, concentration 1.0 mol dm^{-3}

(ii) E.m.f. is $(+0.34) - (-0.76) = +1.10$ V.

(iii) A very high internal resistance voltmeter.

(iv) From the zinc to the copper.

(v) $Zn(s) + Cu^{2+}(aq) \rightarrow Zn^{2+}(aq) + Cu(s)$.

Ion–electron half-equation	$E^{\ominus}/V$
$\frac{1}{2}I_2 + 2e^- \rightleftharpoons I^-$	$+0.54$
$Fe^{3+} + e^- \rightleftharpoons Fe^{2+}$	$+0.77$
$Ag^+ + e^- \rightleftharpoons Ag$	$+0.80$
$BrO_3^- + 6H^+ + 5e^- \rightleftharpoons \frac{1}{2}Br_2 + 3H_2O$	$+1.52$

oxidizing power

reducing power

(c) (i) Bromate(V) ions would oxidize iodide ions to iodine molecules and be reduced to bromine molecules, $BrO_3^- + 6H^+ + 5I^- \rightarrow \frac{1}{2}Br_2 + 3H_2O + 2\frac{1}{2}I_2$, because the E for the acidified bromate is more positive than that for iodine.

(ii) No reaction because the E for silver ions is more positive than that for iron(III) ions. (Silver ions would oxidize iron(II) ions to iron(III) ions and be reduced to silver.)

(iii) iron(III) ions would oxidize iodide ions to iodine molecules and be reduced to iron(II) ions, $Fe^{3+} + I^- \rightarrow Fe^{2+} + \frac{1}{2}I_2$, because the E for iron(III) ions is more positive than that for iodine.

Energetics II Free Energy and Entropy

In this chapter you will meet the idea of a free energy, G, and entropy, S, as indicators of the feasibility of a reaction. When you see the symbols $\Delta G^{\ominus}$ and $\Delta S^{\ominus}$ you will know they refers to changes under standard conditions of temperature, pressure, etc. For A-level you should know how to calculate changes in free energy and entropy and to use the results to predict whether or not a reaction would be spontaneous. Master this section. It will help you achieve the top grade.

By the end of this chapter you will be able to

→ Calculate the free energy change for a reaction from free energy changes of formation of reactants and products

→ Calculate free energy change from $\Delta G^{\ominus} = -RT\ln K$ and from $\Delta G^{\ominus} = -zFE^{\ominus}$

→ Calculate the entropy change for a reaction from the entropies of reactants and products

→ Draw and use Ellingham diagrams

Topic checklist

Tick each of the boxes below when you are satisfied that you have mastered the topic

	Edexcel		AQA		OCR		WJEC		CCEA	
	AS	A2	AS	A2	AS	A2	AS	A2	AS	A2
Free energy changes		●		●		●		●		●
Entropy		●		●		●		●		●

Free energy changes

In theory all chemical reactions are reversible and may come to equilibrium. The more the energetic stablility of the products exceeds that of the reactants, the higher the equilibrium constant, K, and the more complete the reaction will be. The free energy change, ΔG, is related to K and indicates the feasibility of a reaction.

Energy changes

Enthalpy change, ΔH

→ Most reactions are exothermic (ΔH negative) and feasible because the products are energetically more stable than the reactants.

→ Most endothermic (ΔH positive) reactions are not feasible because the reactants are energetically more stable than the products.

You should be able to calculate the standard molar enthalpy change for a reaction, ΔH_r, °from ΔH_f values of the reactants and products. E.g.:

$$H^+(aq) + HCO_3^-(aq) \rightarrow H_2O(l) + CO_2(g)$$

$\Delta H_f^\ominus \quad\quad 0 \quad\quad\quad -692 \quad\quad\quad -286 \quad\quad -394 \text{ kJ mol}^{-1}$

so $\Delta H_r^\ominus$ is $[(-286) + (-394)] - [(0) + (-692)] = +12$ kJ mol^{-1}.

This reaction of acid on hydrogencarbonate is endothermic and would not seem energetically feasible but it works!

→ H is not always an accurate measure of energetic feasibility.

Free energy change, ΔG

ΔG is a thermodynamic quantity related to ΔH. You can find values for the standard free energy change of formation, ΔG_f, alongside ΔH_f values in data books. You could use ΔG_f, values in the same way as ΔH_f values to calculate the standard free energy change, ΔG_r, for the reaction above.

$$H^+(aq) + HCO_3^-(aq) \rightarrow H_2O(l) + CO_2(g)$$

$\Delta G_f^\ominus \quad\quad 0 \quad\quad\quad -587 \quad\quad\quad -286 \quad\quad -394 \text{ kJ mol}^{-1}$

so $\Delta G_r^\ominus$ is $[(-286) + (-394)] - [(0) + (-587)] = -93$ kJ mol^{-1}.

→ A reaction will be energetically feasible (spontaneous) if its standard molar free energy change, ΔG_r, °is negative.

Free energy change and ln$K$

You can calculate the free energy change of a reaction from its equilibrium constant using the following relationship.

$$\Delta G_T^\ominus = -RT\ln K \text{ or } \Delta G_T^\ominus = -2.303RT\log K$$

For example,

Chemical system at 500 K	$\Delta H_{500}^\ominus$/kJ mol^{-1}	K_p	$\Delta G_{500}^\ominus$/kJmol^{-1}
$H_2(g) + CO_2(g) \Leftrightarrow H_2O(g) + CO(g)$	+41	7.8×10^{-3}	+20
$N_2(g) + 3H_2(g) \Leftrightarrow 2NH_3(g)$	−101	3.8×10^{-2}/atm^{-2}	+14
$N_2O_4(g) \Leftrightarrow 2NO_2(g)$	+57	1.7×10^3/atm	-31
$2SO_2(g) + O_2(g) \Leftrightarrow 2SO_3(g)$	-200	2.5×10^{10}/atm^{-1}	-90

The jargon

Stable kinetically = very high E_a (activation energy) value.
A reaction is *spontaneous* if the reactants are energetically unstable with respect to the products <u>even if the reaction mixture is kinetically stable</u>.
Not feasible energetically = (free) energy of products > (free) energy of reactants. ΔG is positive.

Action point

Make a list of feasible endothermic processes including evaporation of liquids and decompression of gases.

The jargon

G = Gibbs free energy after J. Willard Gibbs, Professor of Mathematical Physics at Yale University in 1875.

Checkpoint 1

Calculate $\Delta H^\ominus$ and $\Delta G^\ominus$ for the process $KI_{(s)} \rightarrow KI_{(aq)}$ from the following data and comment on the feasibility of dissolving potassium iodide in water.

	$\Delta H^\ominus_f$	$\Delta G^\ominus_f$
$KI_{(s)}$	-328 kJ mol^{-1}	-325 kJ mol^{-1}
$KI_{(aq)}$	-307 kJ mol^{-1}	-335 kJ m ol^{-1}

Watch out!

The value of R (8.314) is in J K^{-1} mol^{-1} but the value of ΔG_T is in kJ K^{-1} mol^{-1}.

Checkpoint 2

For the reaction $X(g) \leftrightarrows Y(g)$, what would be (a) the value of K if $\Delta G = 0$ and (b) the percentage conversion of gas X into gas Y?

→ As K increases the molar free energy change, ΔG_r, decreases and the reaction becomes more energetically feasible (spontaneous).

Free energy change and E_{cell}

You can calculate the free energy change of an electrochemical cell reaction from its $E^{\ominus}_{cell}$ value using the following relationship.

$$\Delta G^{\ominus} = -nFE^{\ominus}_{cell}$$

where z is the number of moles of electrons transferred per mole of reaction specified by the equation for the cell reaction and F is the Faraday constant = 96 500 C mol^{-1}.

For example, in the cell

$$Zn(s) \mid Zn^{2+}(aq) \vdots\vdots Pb^{2+}(aq) \mid Pb(s) \; E^{\ominus}_{cell} = +0.63 \text{ V}$$

the spontaneous redox reaction is

$$Zn(s) + Pb^{2+}(aq) \rightarrow Zn^{2+}(aq) + Pb(s)$$

$$
\begin{aligned}
\Delta G^{\ominus} \quad &= -nFE^{\ominus}_{cell} \\
&= -2 \times 96\,500 \times 0.63 \\
&= -121\,590 \text{ J mol}^{-1} \\
&= -122 \text{ kJ mol}^{-1}.
\end{aligned}
$$

ln$K$ and $E^{\ominus}_{cell}$

The standard electrode potential of a half cell is its emf measured against the standard hydrogen electrode, e.g.:

$$Zn^{2+}(aq) + 2e^- \rightarrow Zn(s) \; E^{\ominus} = -0.76\text{V}$$

We can combine half cells and calculate the standard emf of the from the half cell potentials, e.g.:

$$Zn(s) \mid Zn^{2+}(aq) \vdots\vdots Cu^{2+}(aq) \mid Cu(s); \; E^{\ominus}_{cell} = +1.10 \text{ V}$$

Now from $\Delta G_T^{\ominus} = -RT\ln K$ (or $\Delta G_T^{\ominus} = -2.303RT\log K$)

and $\Delta G^{\ominus} = -zFE^{\ominus}_{cell}$

it follows that $-RT\ln K$ ($-2.303RT\log K$) $= -zFE^{\ominus}_{cell}$

so $\ln K = (zF/RT)E^{\ominus}_{cell}$ and from this expression we can calculate the equilibrium constant for the cell reaction, which in our example is

$$Zn(s) + Cu^{2+}(aq) \rightarrow Zn^{2+}(aq) + Cu(s)$$

$$
\begin{aligned}
\ln K \quad &= (2 \times 96\,500/8.314 \times 298) \times 1.10 \\
&= 85.69
\end{aligned}
$$

So $K = 1.6 \times 10^{37}$

Exam practice (12 minutes)

(a) Use the following values for the enthalpies and free energies of formation to calculate the enthalpy and free energy changes for the reaction of ammonia with (i) hydrogen chloride and (ii) hydrogen bromide

	NH$_3$	HCl	HBr	NH$_4$Cl	NH$_4$Br
ΔH/kJmol^{-1}	-46.1	-92.3	-36.4	-314	-527
ΔG/kJmol^{-1}	-16.5	-95.2	-53.4	-203	-488

(b) (i) Name the type of bond formed when ammonia combines with a hydrogen halide,

(b) (ii) comment on the spontaneity of the reactions in (a) and

(b) (iii) suggest a reason for the difference in the ΔG values of the two ammonium halides.

Watch out!

Coulomb x Volt = Joule (NOT kilojoule)

Don't forget!

Many reactions which are predicted from thermodynamic data to be feasible do not take place. Thermodynamics cannot predict kinetic data. If the activation energy is too high then an energetically feasible reaction will not take place a measurable rate.

Jargon

Standard electrode potentials are sometimes called standard reduction potentials because the forward reaction is a reduction.

Checkpoint 3

(a) State two ways of measuring the emf of a cell.
(b) What is the underlying principle behind these methods?

Watch out!

The standard electrode potential does is not doubled if you double the number of moles taking part in a cell reaction.

Checkpoint 4

Calculate the value of $E^{\ominus}_{cell}$ for
$Fe(s) \mid Fe^{2+}(aq) \vdots\vdots Ag^+(aq) \mid Ag(s)$
You must look up the standard electrode potentials for
$Ag^+(aq) + e^- \rightarrow Ag(s)$ and
$Fe^{2+}(aq) + 2e^- \rightarrow Fe(s)$

Grade booster

If you want the top grade be prepared to predict the feasibility of a reaction from given values of standard electrode potentials (see page 76).

Entropy

Entropy is a measure of the random dispersal of energy of a system. Changes in entropy can be related to changes in enthalpy and free energy.

The jargon

If you want the top grade do not confuse entropy, symbol S, with enthalpy, symbol H. Entropy is related to the disorder of a system. If a system becomes more disordered then entropy increases.

Don't forget

Free energy may be seen as the balance between enthalpy and entropy that determines a reaction's feasibility.

Action point

Make a list of three endothermic processes which occur spontaneously and in which entropy increases.

Checkpoint 1

(a) Calculate the change in entropy, ΔS, for the reaction of ammonia with hydrogen iodide using the following standard molar entropy values in J mol^{-1} K^{-1}.

NH$_3$	HI	NH$_4$I
192	207	117

(b) State whether the entropy decreases or increases and suggest an explanation for the change.

"The total entropy always increases if a spontaneous reaction occurs"

Second Law Of Thermodynamics

Don't forget

Many energetically feasible reactions do not take place under ordinary conditions because the activation energy, E_a, is too high. ΔG, ΔH and ΔS tell us nothing about the rate of a reaction.

Entropy changes

Physical changes

When water is in its solid state (as ice) the molecules, held by hydrogen bonding in an open crystal lattice structure, have vibrational energy. When the ice melts to a liquid, the molecules gain rotational and translational energy.

The entropy of the water increases: $H_2O(s) \rightarrow H_2O(l)$; $\Delta S^\ominus = +25.1$ J mol^{-1} K^{-1}

The entropy increases still further when liquid water turns into a vapour (the gaseous state): $H_2O(l) \rightarrow H_2O(g)$; $\Delta S^\ominus = +118$ J mol^{-1} K^{-1}.

Chemical changes

We can calculate the standard entropy change, $\Delta S^\ominus$, for a reaction from the standard enthalpies, S, of the reactants and products:

$$\Delta S^\ominus = \Sigma S^\ominus_{products} - \Sigma S^\ominus_{reactants}$$

For example, $CoCl_2.6H_2O(s) + 6SOCl_2(l) \rightarrow CoCl_2(s) + 12HCl(g) + 6SO_2(g)$

$S^\ominus$ / J mol^{-1} K^{-1} 343 6×308 109 $12 \times 1876 \times 248$

$\Delta S^\ominus$ is $3841 - 2191 = +1650$ J mol^{-1} K^{-1} (= 1.65 kJ mol^{-1} K^{-1})

→ Entropy increases significantly when gases are produced.

Entropy and free energy

We may think of the chemicals (reactants and products) in a test tube as a system and everything else (the test tube, air in the laboratory, etc.) as their surroundings.

According to thermodynamics, $\Delta S_{total} = -\Delta G/T$ and the total entropy always increases if a spontaneous reaction occurs.

So it follows that ΔG (= $-T \times \Delta S_{total}$) must be negative for a reaction to be spontaneous (energetically feasible).

The total entropy change is the sum of the change in entropy of the system and the surroundings,

$$\Delta S_{total} = \Delta S_{surroundings} + \Delta S_{system}$$

According to thermodynamics, $\Delta S_{surroundings} = -\Delta H/T$

Substituting for ΔS_{total} and $\Delta S_{surroundings}$ gives

$$(-\Delta G/T) = (-\Delta H/T) + \Delta S_{system}$$

Rearranging this equation gives

$$\Delta G = \Delta H - T \times \Delta S$$

where ΔG, ΔH and $\times \Delta S$ represent the *values at temperature T*. We often indicate this by writing the equation in the following way,

$$\Delta G_T = \Delta H_T - T \times \Delta S_T,$$

→ $\Delta G^\ominus = \Delta H^\ominus - T \times \Delta S^\ominus$ where T is the standard temperature (usually 298K) and the values of the changes in free energy, enthalpy and total entropy are the standard values at the standard temperature T.

Conditions for feasible reactions

If we think of a free energy change as the balance between an enthalpy change and an entropy change that decides the feasibility of a process then we may consider what combination of ΔH and ΔS values will make ΔG negative (i.e. ΔS_{total} positive).

$\Delta H^{\ominus}$	$\Delta S^{\ominus}$	$\Delta G^{\ominus}$	$= \Delta H^{\ominus} - T \times \Delta S^{\ominus}$		Feasible?
–	+	Negative			Yes – always
–	–	Negative if	$\|T \times \Delta S^{\ominus}\| <$	$\|\Delta H^{\ominus}\|$	Maybe
		Positive if	$\|T \times \Delta S^{\ominus}\| >$	$\|\Delta H^{\ominus}\|$	
+	+	Negative if	$\|T \times \Delta S^{\ominus}\| >$	$\|\Delta H^{\ominus}\|$	Maybe
		Positive if	$\|T \times \Delta S^{\ominus}\| <$	$\|\Delta H^{\ominus}\|$	
+	–	Positive			No – never

→ A reaction is feasible if $\Delta G^{\ominus}$ is negative (ΔS_{total} is positive).

What makes reduction of oxides by carbon feasible?

An Ellingham diagram shows $\Delta G_f^{\ominus}$ of oxides against T.

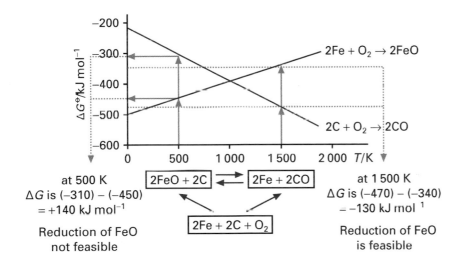

at 500 K
ΔG is $(-310) - (-450)$
$= +140$ kJ mol^{-1}

Reduction of FeO
not feasible

at 1 500 K
ΔG is $(-470) - (-340)$
$= -130$ kJ mol^{-1}

Reduction of FeO
is feasible

What makes electrochemical reactions feasible?

For the electrochemical cell Pb(s) | Pb^{2+}(aq) Zn2+(aq) | Zn(s), $E^{\ominus}_{cell} =$ −0.63 V and the reaction for the cell as written is Pb(s) + Zn^{2+}(aq) → Pb^{2+}(aq) + Zn(s)
$\Delta G^{\ominus} = nFE^{\ominus}_{cell} = -2 \times 96\,500 \times (-0.63) = +122\,000$ J mol^{-1}, ($+122$ kJ mol^{-1}) so the reaction is **not** feasible. $\Delta G^{\ominus}$ for the reverse reaction is -122 kJ mol^{-1} and E_{cell} for the cell written the other way around is $+0.63$ V.

→ Zinc displaces lead from its aqueous lead(II) cations and the feasible reaction is Zn(s) + Pb^{2+}(aq) → Zn^{2+}(aq) + Pb(s)

Exam practice (10 minutes)   answers: page 87

Discuss the following in terms of changes in free energy and entropy.

(a) The reaction of sodium hydrogencarbonate with acid is endothermic but feasible.

(b) Barium chloride is soluble in water but barium sulphate is insoluble.

(c) The reduction Al$_2$O$_3$ + 3C → 2Al + 3CO is feasible above 2100 K but the reduction Al$_2$O$_3$ + 3CO → 2Al + 3CO$_2$ is not feasible at any temperature.

Jargon

|y| means the magnitude or numerical value of y ignoring the + or - sign.

Watch out!

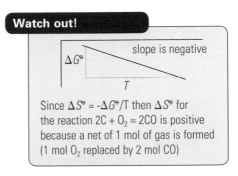

Since $\Delta S^{\ominus} = -\Delta G^{\ominus}/T$ then $\Delta S^{\ominus}$ for the reaction 2C + O$_2$ = 2CO is positive because a net of 1 mol of gas is formed (1 mol O$_2$ replaced by 2 mol CO)

Action point

Use Ellingham diagrams (from a data book or the Web) to sort aluminium, chromium, copper, lead, magnesium, silver and zinc into a table starting with the metal with the most stable oxide at 0 °C. Extend the table with one major industrial use for each metal. Where would you place gold and iron in the table?

Checkpoint 2

(a) Calculate the change in entropy for the decomposition of dinitrogen oxide to nitrogen and oxygen using $\Delta G = \Delta H - T \times \Delta S$, and the following values, in kJ mol^{-1}, for the formation of N$_2$O at 298K.
$\Delta G_f = +104$ and $\Delta H_f = +82$.

(b) State whether the entropy decreases or increases and suggest an explanation for the change.

85

Structured exam question

answers: page 87

(a) State concisely in words what the following symbols represent

(i) $\Delta H^{\ominus}$ ....
....

(ii) ΔG_T ....
....

(iii) $\Delta S^{\ominus}$ ....
.... [3 marks]

(b) Write an equation to show how the three changes in (a) above are related to the Kelvin temperature, T.

.... [1 mark]

(c) (i) Write an expression for the equilibrium constant, K_p, for the reaction $SF_4(g) + F_2(g) \Leftrightarrow SF_6(g)$ and state its units.

....[2 marks]

(c) (ii) Write an expression relating $\Delta G^{\ominus}$ to the equilibrium constant, K_p,

....[1 mark]

(d) (i) A Data Book gives the following information on sulphur tetrafluoride, sulphur hexafluoride and fluorine.

	$\Delta H_f^{\ominus}$/kJ mol^{-1}	$\Delta S^{\ominus}$/J mol^{-1} K^{-1}
fluorine, $F_2(g)$	0	158.6
sulphur tetrafluoride, $SF_4(g)$	-774.9	291.9
sulphur hexafluoride, $SF_6(g)$	-1209	291.7

Use the equation in (b) above to calculate the value of $\Delta G^{\ominus}$ at 25°C for the reaction $SF_4(g) + F_2(g) \rightarrow SF_6(g)$

....
....
.... [3 marks]

(d) (ii) Use the expression in (c)(i) and the value of $\Delta G^{\ominus}$ in (d)(i) to calculate the equilibrium constant for the reaction $SF_4(g) + F_2(g) \rightleftharpoons SF_6(g)$ at 25°C.

....
....
....
....
....
....

[2 marks]

Answers
Energetics II

Free energy changes

Checkpoints

1 ΔH°_f is $(-307) - (-328) = +21$ kJ mol^{-1}

ΔG°_f is $(-335) - (-325) = -10$ kJ mol^{-1}

The free energy change is negative so dissolving potassium iodide in water is energetically feasible. Since the enthalpy change is positive the salt will dissolve endothermically.

2(a) $-RT\ln K = 0$ R and T are not zero, so $\ln K = 0$ and $K = 1$

2(b) If $K = 1$ then $[Y(g)]/[X(g)] = 1$ so 50% X(g) is converted into Y(g).

3(a) Use a very high internal resistance voltmeter or a potentiometer circuit

3(b) To find the voltage when the cell is delivering zero current.

4 E_{cell} is $(+ 0.80) - (-0.44) = +1.24$ volts

Exam practice

(a) $NH_3 + HCl \rightarrow NH_4Cl$

½N$_2$ + 2H$_2$ + ½Cl$_2$

$(-46.1) + (-92.3) + \Delta H = (-314)$

so ΔH is $(-314) - (-46.1) - (-92.3) = -176$ kJ mol^{-1}

Similarly

ΔG is $(-203) - (-16.5) - (-95.2) = -91.3$ kJ mol^{-1}

$NH_3 + HBr \rightarrow NH4Br$

½N$_2$ + 2H$_2$ + ½Br$_2$

$(-46.1) + (-36.4) + \Delta H = (-527)$

so ΔH is $(-527) - (-46.1) - (-36.4) = -444$ kJ mol^{-1}

Similarly

ΔG is $(-488) - (-16.5) - (-53.4) = -418$ kJ mol^{-1}

(b)(i) dative covalent or coordinate bond between the ammonia molecule and the hydrogen

(b)(ii) both exothermic reactions are energetically feasible because their free energy changes are negative

(b)(iii) the H–Br bond is weaker than the H–Cl bond allowing the ammonia more readily to remove the proton to form the

Entropy

Checkpoints

1(a) ΔS is $117 - (192 + 207) = -282$ J mol^{-1} K^{-1}.

1(b) Entropy decreases. Two moles of gas (ammonia and hydrogen iodide) are replaced by on mole of solid (ammonium iodide)

2(a) $N_2 + \frac{1}{2}O_2 \rightarrow N_2O$; $\Delta H_f = +82$ kJ mol^{-1}; $\Delta G_f = +104$ kJ mol^{-1}, for the decomposition (the reverse) the values are negative.

$(-104) = (-82) - 298 \times \Delta S$

so $298 \times \Delta S$ is $(-82) - (-104) = +22$

ΔS is $+22/298 = 0.0738$ kJ mol^{-1}; 73.8 J mol^{-1}.

2(b) Entropy increases. One moles of gas (dinitrogen oxide) is replaced by one and a half moles of gas (1 mole nitrogen and ½ mole of oxygen).

Exam practice

(a) ΔG is negative because ΔS is highly positive when the solid forms a solution and gives off CO_2 gas. These products are more disordered than the reactants and the energy distribution more varied.

(b) The difference in the lattice (free) energy and the hydration (free) energies for the barium salts gives ΔG a negative value for barium chloride but a positive value for barium sulphate. The difference in the ion size and the charge of the chloride and sulphate ions determines the difference in the lattice and hydration energies of the salts.

(c) Above 2100K, ΔG for $3C + 1\frac{1}{2}O_2 \rightarrow 3CO$ is more negative (ΔS positive) and therefore more feasible than $2Al + 1\frac{1}{2}O_2 \rightarrow Al_2O_3$. The formation of Al_2O_3 is always more negative and therefore more feasible than $3CO + 1\frac{1}{2}O_2 \rightarrow 3CO_2$ (ΔS negative) at any temperature.

Structured Exam practice

(a)(i) standard enthalpy change (at 298 kelvin)

(a)(ii) standard free energy change at temp T

(a)(iii) standard (absolute) entropy (at 298 kelvin)

(b) $\Delta G_T = \Delta H_T - T \times \Delta S$

(c)(i) $K_p = P_{SF_6}/(P_{SF_4} \times P_{F_2})$ units of 1/pressure (e.g. atm^{-1})

(c)(ii) $\Delta G^\circ = -RT\ln K$

(d)(i) $\Delta G_{298} = (-1209)-(-774.9) - 298\times\{(291.7)-(291.9+158.6)\}/1000$

$= (-434.1) - 298 \times (-158.8)/1000$

$= (-434.1) + 47.32$

$= -386.8$ kJ mol^{-1}

(d)(ii) $\Delta G_{298} = -RT\ln K$

$-386.8 = -(8.314/1000) \times 298 \times \ln K$

$\ln K = 386.8/2.478$

$= 156.1$

$K = 6.042 \times 10^{67}$

Inorganic chemistry

You must get used to the periodic table but there is no need to learn it by heart. See how elements are organized into groups with characteristic properties. Notice the trends in properties down the table (within each group) and across the table (from group I to group VII). Between groups II and III you should see that the series of d-block elements across the table shows more similarities than trends. Do make sure you check to see which aspects of inorganic chemistry are included in your particular syllabus.

By the end of this chapter you will be able to

- Describe and explain the trend in the following properties of elements going across a period: 1st ionization energy, atomic radius, melting point, electronegativity
- Describe the bonding and reactions of the chlorides or the oxides across a period and the acid–base nature of the oxides across a period
- Describe and explain the characteristics, reactions and trends in reactivity of groups I to VIII (or 0) and the transition metals

Topic checklist

	Edexcel		AQA		OCR		WJEC		CCEA	
	AS	A2	AS	A2	AS	A2	AS	A2	AS	A2
The periodic table	○		○		○		○		○	
Patterns and trends in the periodic table	○		○		○		○		○	
Groups I and II: alkali and alkaline earth metals	○		○		○		○		○	
Industrial chemistry: s-block		●		●		●		●		●
Group III: aluminium and boron		●		●		●		●		●
Group IV: elements and oxides		●		●		●		●		●
Group IV: chlorides and hydrides		●		●		●		●		●
Group V: elements and oxides		●		●		●		●		●
Group V: oxoacids and hydrides		●		●		●		●		●
Group VI: oxygen and sulphur		●		●		●		●		●
Group VI: water and hydrogen peroxide		●		●		●		●		●
Group VII: halogens and hydrogen halides	○		○		○		○		○	
Group VII: halides and interhalogen compounds	○		○		○		○		○	
Industrial chemistry: aluminium and carbon	○	●	○	●	○	●	○	●	○	●
Industrial chemistry: silicon and nitrogen	○		○		○		○		○	
Industrial chemistry: sulphur and the halogens	○		○		○		○		○	
Group VIII (or 0): the noble gases		●		●		●		●		●
1st transition series: metals, aqueous ions and redox		●		●		●		●		●
1st transition series: redox reactions and complex ion formation		●		●		●		●		●
1st transition series: chromium		●		●				●		●
1st transition series: manganese		●						●		
1st transition series: copper		●				●		●		●

Tick each of the boxes below when you are satisfied that you have mastered the topic

The periodic table

Action point

Draw on the periodic table a line which divides metals from non-metals.

s-block

I II

d-block ⟶

1.01 **H** Hydrogen 1								
6.94 **Li** Lithium 3	9.01 **Be** Beryllium 4							
23.0 **Na** Sodium 11	24.3 **Mg** Magnesium 12							
39.1 **K** Potassium 19	40.1 **Ca** Calcium 20	45.0 **Sc** Scandium 21	47.9 **Ti** Titanium 22	50.9 **V** Vanadium 23	52.0 **Cr** Chromium 24	54.9 **Mn** Manganese 25	55.8 **Fe** Iron 26	58.9 **Co** Cabalt 27
85.5 **Rb** Rubidium 37	87.6 **Sr** Strontium 38	88.9 **Y** Yttrium 39	92.9 **Zr** Zirconium 40	92.9 **Nb** Niobium 41	95.9 **Mo** Molybdenum 42	98.9 **Tc** Technetium 43	101.1 **Ru** Ruthenium 44	103 **Rh** Rhodium 45
133 **Cs** Caesium 55	137 **Ba** Barium 56	139 **La** Lanthanum 57	179 **Hf** Hafnium 72	181 **Ta** Tantalum 73	184 **W** Tungsten 74	186 **Re** Rhenium 75	190 **Os** Osmium 76	192 **Ir** Iridium 77
(223) **Fr** Francium 87	(226) **Ra** Radium 88	(227) **Ac** Actinium 24						

f-block ⟶

140 **Ce** Cerium 58	141 **Pr** Praseodymium 59	144 **Nd** Neodymium 60	(147) **Pm** Promethium 61	150 **Sm** Samarium 62	(153) **Eu** Europium 63
232.0 **Th** Thorium 90	(231) **Pa** Protoactinium 91	238.1 **U** Uranium 92	(237) **NP** Neptunium 93	(244) **Pu** Plutonium 94	(243) **Am** Americium 95

Action point

(a) Describe how the acid–base nature of oxides varies across the period Na to Cl.
(b) State how the nature of the bonding in the chlorides varies from Na to S and illustrate this by describing how at least *one* chloride for *each* element behaves when treated with water.
(c) Compare and explain the difference in the reaction of CCl_4 and $SiCl_4$ with water.

Key

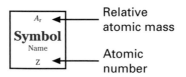

A_r — Relative atomic mass

Symbol Name

Z — Atomic number

p-block

	III	IV	V	VI	VII	4.00 **He** Helium 2
	10.8 **B** Boron 5	12.0 **C** Carbon 6	14.0 **N** Nitrogen 7	16.0 **O** Oxygen 8	19.0 **F** Fluorine 9	20.2 **Ne** Neon 10
→	27.0 **Al** Aluminium 13	28.1 **Si** Silicon 14	31.0 **P** Phosphorus 15	32.1 **S** Sulphur 16	35.5 **Cl** Chlorine 17	40.0 **Ar** Argon 18

58.7 **Ni** Nickel 28	63.5 **Cu** Copper 29	65.4 **Zn** Zinc 30	69.7 **Ga** Gallium 31	72.6 **Ge** Germanium 32	74.9 **As** Arsenic 33	79.0 **Se** Selenium 34	79.9 **Br** Bromine 35	83.8 **Kr** Krypton 36
106 **Pd** Palladium 46	108 **Ag** Silver 47	112 **Cd** Cadmium 48	115 **In** Indium 49	119 **Sn** Tin 50	122 **Sb** Antimony 51	128 **Te** Tellurium 52	127 **I** Iodine 53	131 **Xe** Xenon 54
195 **Pt** Platinum 78	197 **Au** Gold 79	201 **Hg** Mercury 80	204 **Tl** Thallium 81	207 **Pb** Lead 82	209 **Bi** Bismuth 83	(210) **Po** Polonium 84	(210) **At** Astatine 85	(222) **Rn** Radon 86

→

157 **Gd** Gadolinium 64	159 **Tb** Terbium 65	162 **Dy** Dysprosium 66	165 **Ho** Holmium 67	167 **Er** Erbium 68	169 **Tm** Thulium 69	173 **Yb** Ytterbium 70	175 **Lu** Lutetium 71
(247) **Cm** Curium 96	(245) **Bk** Berkelium 97	(251) **Cf** Californium 98	(254) **Es** Einsteinium 99	(253) **Fm** Fermium 100	(256) **Md** Mendelevium 101	(254) **No** Nobelium 102	(257) **Lr** Lawrencium 103

Action point

Show on the periodic table the most electronegative element and the least electronegative element. Indicate which of the elements form compounds which show hydrogen bonding.

Check the net

You can download a copy of the periodic table for your own use from www.webelements.com.

You will need to have Adobe Acrobat Reader installed.
This can be downloaded free of charge from www.adobe.com.

Patterns and trends in the periodic table

In 1869 Dmitri Mendeléev arranged elements in a table and predicted the existence and properties of two elements that were later discovered. Since then a periodic table in one form or another has been a cornerstone of chemistry.

Periodicity

→ Similar properties recur at regular intervals when the elements are arranged in order of increasing atomic number.
→ Similarity of properties of the elements and their compounds occur within each group of the s- and p-blocks.

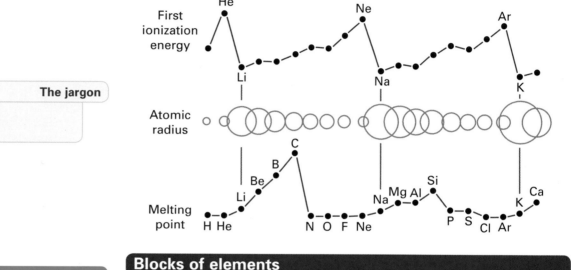

Blocks of elements

At.no. 1 = Hydrogen
At.no. 2 = Helium
1st period = Li to Ne
s-block = Groups I and II
p-block = Groups III–VIII
d-block = Transition elements
f-block = Lanthanides and Actinides

→ Most elements are metals.
→ Group I elements are the most reactive metals.
→ Group VII elements are the most reactive non-metals.

Atypical properties

Each element in the first period (Li to F) has some properties not typical of their group.

Group I	Decomposition of $Li_2CO_3 \rightarrow Li_2O + CO_2$ on heating in a test tube. The other alkali metal carbonates do not decompose.
Group IV	Tetrachloromethane does not hydrolyze in water. The other tetrachloro compounds do, e.g. $SiCl_4 + 2H_2O \rightarrow SiO_2 + 4HCl$.
Group VII	Silver fluoride dissolves in water but the other silver halides are insoluble, e.g. $Ag^+(aq) + Cl^-(aq) \rightarrow AgCl(s)$.

Links

See page 94: lithium; 98: boron; 112: fluorine.

Trends in properties

You need to know the important broad trends in the properties of elements and their compounds

→ reactivity of the s-block metals *increases* with increasing atomic number down each group
→ reactivity of the halogens and group VI elements *decreases* with increasing atomic number down each group
→ character of group IV elements changes from non-metal to metal with increasing atomic number down the group

From left to right across the s- and p-block elements

→ atomic and ionic radius decreases
→ tendency of chlorides to hydrolyze increases

Within the series of transition elements

→ metallic character shows little or no change with increasing atomic number across the series
→ basic character decreases and acidic character increases as the oxidation number increases

+2 basic   +3 amphoteric   +6 acidic

Fe^{2+} Cr^{2+} Fe^{3+} Cr^{3+}

FeO_2^- $[Cr(OH)]_6^{3-}$ FeO_4^2 $Cr_2O_7^{2-}$

Some important patterns and trends for the s- and p-blocks are summarized below:

Increase →

Elements	*Elements*
Atomic radius	First ionization energy
Ionic radius	Electronegativity
Reducing power	Oxidizing power
Oxides	*Oxides*
Basic character	Acidic character
Oxides, hydrides, chlorides	*Oxides, hydrides, chlorides*
Ionic character	Covalent character

Increase ↓   Increase ↑

Exam practice (5 minutes)   answers: page 135

(a) Give one example of each of the following
   (i)   an amphoteric oxide
   (ii) a semiconductor element
   (iii) a group IV oxide which is an oxidizing agent

(b) Give *two* trends in chemical properties, with increasing atomic number, for the elements of group IV.

Groups I and II: alkali and alkaline earth metals

The Group I elements are the most reactive metals in the Periodic Table. The Group II elements are less reactive and, except for beryllium, more like typical metals.

The elements

Trends in physical properties

	Electronic configuration	M.p./°C	$\rho/\mathrm{g\,cm^{-3}}$	$E_{ml}/\mathrm{kJ\,mol^{-1}}$	
Li	$1s^2 2s$	180	0.53	519	
Na	$1s^2 2s^2 2p^6 3s$	97.8	0.97	494	
K	$1s^2 2s^2 2p^6 3s^2 3p^6 4s$	63.7	0.86	418	
Rb	$1s^2 2s^2 2p^6 3s^2 3p^6 3d^{10} 4s^2 4p^6 5s$	39.9	1.53	402	
Cs	[krypton core]$4d^{10} 5s^2 5p^6 6s$	28.7	1.54	376	
Be	$1s^2 2s^2$	1278	1.85	900	hcp
Mg	$1s^2 2s^2 2p^6 3s^2$	649	1.74	736	hcp
Ca	$1s^2 2s^2 2p^6 3s^2 3p^6 4s^2$	839	1.54	590	fcc
Sr	$1s^2 2s^2 2p^6 3s^2 3p^6 3d^{10} 4s^2 4p^6 5s^2$	769	2.62	548	fcc
Ba	[krypton core]$4d^{10} 5s^2 5p^6 6s^2$	725	3.51	502	bcc

→ All alkali metals are soft and have a body-centred cubic structure.
→ All s-block elements are ductile, malleable and conduct electricity.

Trends in chemical properties

→ All alkali metals react with water with increasing violence down the group (Li–Cs), e.g. $2Na(s) + 2H_2O(l) \rightarrow 2NaOH(aq) + H_2(g)$.
→ Magnesium burns in water vapour to form magnesium oxide and the rest of the group (Ca–Ba) react with water to form hydroxides.
→ s-Block elements combine with halogens and oxygen to form halides, oxides, peroxides and superoxides.
→ Magnesium reduces $H^+(aq)$ ion in aqueous acids to $H_2(g)$: the other s-block elements react similarly but too violently for safety.
→ Magnesium reduces the nitrate ion in nitric acid and its aqueous salts to the ammonium ion.

The compounds

Oxides

→ All the oxides are basic except for beryllium oxide which is amphoteric and shows a diagonal relationship with aluminium.
→ Peroxides of sodium (Na_2O_2) and barium (BaO_2) react with water and give off oxygen: $Na_2O_2(s) + H_2O(l) \rightarrow 2NaOH(aq) + \frac{1}{2}O_2(g)$.
→ Potassium, rubidium and caesium form superoxides (KO_2).

Hydroxides

→ Hydroxides form when s-block metals or oxides react with water.
→ All group I hydroxides are water soluble.
→ In group II solubility increases from $Mg(OH)_2$ to $Ba(OH)_2$.

Group II hydroxides are only sparingly soluble in water. Saturated aqueous calcium hydroxide is the 'limewater' which turns milky when you test carbon dioxide: $CO_2(g) + Ca(OH)_2(aq) \rightarrow CaCO_3(s) + H_2O(l)$. The white precipitate 'redissolves' when you pass excess gas into the suspension: $CaCO_3(s) + H_2O(l) + CO_2(g) \rightarrow Ca^{2+}(aq) + 2HCO_3^-(aq)$.

Carbonates

→ Sodium, potassium, rubidium and caesium carbonates are thermally stable and do not decompose on heating in a test tube.
→ You can decompose the carbonates of lithium and the group II metals by heating the solid in a test tube.
→ Thermal stability of group II carbonates increases down the group.
→ Sodium and potassium hydrogencarbonates can exist as solids but calcium hydrogencarbonate exists only in solution.

Nitrates

→ Nitrates are formed when (hydr)oxides react with dilute nitric acid.
→ s-Block nitrates decompose on heating to give off oxygen but lithium and the group II nitrates also give off nitrogen dioxide:
$$2KNO_3(s) \rightarrow 2KNO_2(s) + O_2(g)$$
$$2Ca(NO_3)_2(s) \rightarrow 2CaO(s) + 4NO_2(g) + O_2(g)$$
→ All nitrates are soluble in water.

Sulphates

Calcium sulphate occurs naturally as $CaSO_4$, *anhydrite*, and $CaSO_4 \cdot 2H_2O$, *gypsum*. Magnesium sulphate (Epsom salts) is soluble and crystallizes as $MgSO_4 \cdot 7H_2O$.
→ The solubility of group II sulphates decreases down the group.
   $Ba^{2+}(aq)$ is extremely toxic but $BaSO_4(s)$ is so insoluble that hospital patients swallow a suspension in water for a 'barium meal' X-ray.

Chlorides

→ All chlorides can be formed by direct combination of the elements.
→ All the s-block chlorides are soluble.

Sodium chloride is a major industrial chemical used in the manufacture of sodium hydroxide and chlorine.

Hydrides and nitrides

→ s-Block elements combine with hydrogen to form ionic hydrides.
→ s-Block hydrides react with water to form an alkali and hydrogen.
→ Lithium and magnesium burn in nitrogen to form nitrides.

answers: page 135

Exam practice (6 minutes)

Write an equation for the reaction of (a) water with (i) calcium oxide, (ii) barium peroxide, (iii) lithium hydride; (b) magnesium with (i) nitrogen, (ii) titanium(IV) chloride, (iii) aqueous nitrate ions.

Checkpoint 2

(a) What is (i) hard water and (ii) limescale?
(b) How do stalactites and stalagmites form?
(c) How might the thermal instability of $CaCO_3$ and Li_2CO_3 be related to the polarizing power of the cation?

Examiner's secrets

We often set questions about trends in group II. We are very fond of the trend in the solubilities of the sulphates and hydroxides and the trend in the thermal stabilities of the carbonates.

Don't forget!

We use aqueous barium chloride containing hydrochloric acid to test for sulphate ions:
$Ba^{2+}(aq) + SO_4^{2-}(aq) \rightarrow BaSO_4(s)$ white ppt.

Links

See pages 114–5: halides.

Watch out!

Make sure you remember and understand how and why you use hydrochloric acid when you do flame tests.

Links

See page 15: shapes of molecules.

Industrial chemistry:
s-block

The chemical industry is vital to our economy. Banks have measured prosperity by the tonnage of sulphuric acid manufactured.

The jargon

Brine is concentrated NaCl(aq).
The *diaphragm* is porous asbestos.
The *cell liquor* is NaCl(aq) and NaOH(aq).

Manufacture of sodium hydroxide and chlorine

Checkpoint 1

(a) At which electrodes does (i) oxidation occur, (ii) reduction occur? (b) Explain the function of the diaphragm. (c) Suggest why, other than cost, membrane cells are beginning to replace the Castner–Kellner and the diaphragm cells.

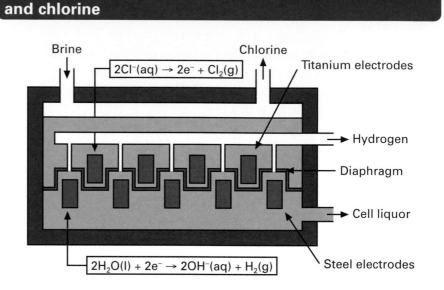

$2Cl^-(aq) \rightarrow 2e^- + Cl_2(g)$

$2H_2O(l) + 2e^- \rightarrow 2OH^-(aq) + H_2(g)$

Brine — Chlorine — Titanium electrodes — Hydrogen — Diaphragm — Cell liquor — Steel electrodes

Checkpoint 2

(a) Combine together and simplify the five equations shown in the Solvay process flow diagram and write the balanced equation for the overall reaction producing sodium carbonate.
(b) Suggest why (i) the plant must be fed with ammonia even though the gas is recycled, and (ii) the Solvay process is being replaced by the mineral trona ($Na_2CO_3 \cdot NaHCO_3 \cdot 2H_2O$) as a source of sodium carbonate.

Manufacture of sodium carbonate

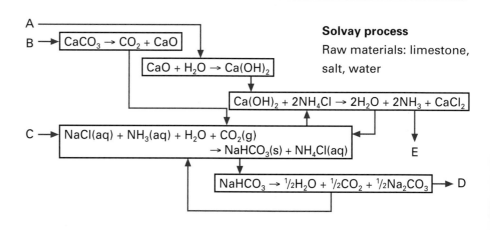

A
B → $CaCO_3 \rightarrow CO_2 + CaO$
$CaO + H_2O \rightarrow Ca(OH)_2$
$Ca(OH)_2 + 2NH_4Cl \rightarrow 2H_2O + 2NH_3 + CaCl_2$
C → $NaCl(aq) + NH_3(aq) + H_2O + CO_2(g) \rightarrow NaHCO_3(s) + NH_4Cl(aq)$
$NaHCO_3 \rightarrow \frac{1}{2}H_2O + \frac{1}{2}CO_2 + \frac{1}{2}Na_2CO_3$ → D
E

Solvay process
Raw materials: limestone, salt, water

➜ In aqueous ammonia (pH ≈ 11), $CO_2(g) + OH^-(aq) \rightleftharpoons HCO_3^-(aq)$.
➜ Sodium hydrogencarbonate is not very soluble in brine.

Links

See page 95: carbonates.

Calcium carbonate and related compounds

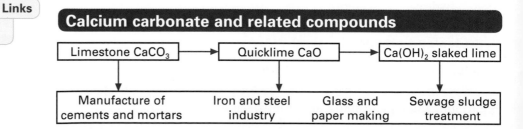

Limestone $CaCO_3$	Quicklime CaO	$Ca(OH)_2$ slaked lime
Manufacture of cements and mortars	Iron and steel industry	Glass and paper making	Sewage sludge treatment

➜ $CaCO_3(s) \rightarrow CaO(s) + CO_2(g)$ in limekilns or in rotary furnaces.
➜ $CaO(s) + H_2O(l) \rightarrow Ca(OH)_2(s)$ in the continuous hydration process.

Milk of lime is an aqueous suspension of calcium hydroxide which, unlike slaked lime, can be pumped easily around the plant.

Extraction of s-block elements

→ s-Block metals have been extracted by electrolysis of molten but *not* aqueous electrolytes.

→ Calcium and magnesium can be extracted from carbonate ores by non-electrolytic methods.

Sodium

at anode:

$$2Cl^- \rightarrow Cl_2 + 2e^-$$

at cathode:

$$2Na^+ + 2e^- \rightarrow 2Na$$

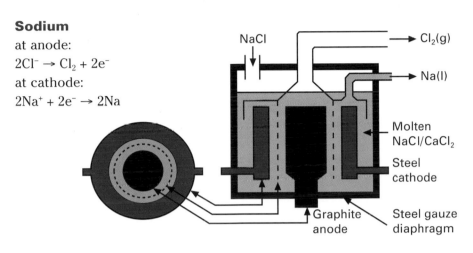

Calcium chloride lowers the melting point of the sodium chloride.

Calcium

The electrolytic extraction of calcium from molten calcium chloride has been replaced by its manufacture from limestone.

1. Limestone is turned into quicklime.
2. Aluminium powder is mixed with quicklime.
3. Briquettes of the mixture are heated under reduced pressure:

$$6CaO(s) + 2Al(s) \rightarrow Ca_3Al_2O_3(s) + 3Ca(g)$$

Magnesium

From seawater:

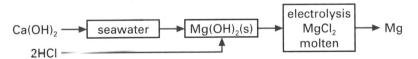

From magnesite (MgO) or dolomite ($CaCO_3 \cdot MgCO_3$):

1. Dolomite is converted into oxide.
2. Iron/silicon alloy is mixed with the oxide.
3. The mixture is heated under reduced pressure to give magnesium vapour.

> **Checkpoint 3**
>
> Suggest why
> (a) the anode is made of graphite and not steel
> (b) the anode is surrounded by a steel gauze diaphragm
> (c) the sodium is not pure.

> **Examiner's secrets**
>
> Check to see which of these industrial processes your Awarding Authority includes in their specification. Even if technical details are not required, the chemistry of the processes may be set in other contexts.

> **Checkpoint 4**
>
> (a) Write an equation for the conversion of limestone into quicklime. (b) What is (i) the function of the aluminium in this reaction, (ii) the oxidation number of the calcium and aluminium before and after reaction?

> **Action point**
>
> Tabulate the industrial products mentioned on these two pages (96 and 97) together with their method of manufacture and a major use for each product.

> **Exam practice (10 minutes)**   answers: page 135
>
> (a) State and explain, with the help of equations, what is observed when (i) a piece of magnesium is heated in air, (ii) one drop of water is added to the residue in a test tube, (iii) a piece of damp red litmus paper is held in the mouth of the test tube.
>
> (b) State one way in which (i) lithium resembles magnesium; (ii) beryllium resembles aluminium.

Group III: aluminium and boron

Aluminium is the most important element of this first group in the p-block. For A-level, boron is unimportant.

Boron

→ Boron *never* forms B^{3+} ions under ordinary conditions.
→ Boron is non-metallic and forms covalent compounds.

The tendency to form covalent compounds is related to the cation polarizing power which increases as the charge on the ion increases and as the radius of the ion decreases.

→ As the first element in group III, boron shows atypical properties because the nuclear charge of its small atom is almost unshielded.

Electron-deficient compounds

BCl_3 and BF_3 are covalent compounds in which the valence shell of the boron atom has only six electrons (*not* eight). Each molecule has three bonding pairs of electrons only but no lone (non-bonding) pair. Such compounds can act as Lewis acids and are called *electron deficient*.

→ BCl_3 and BF_3 are trigonal planar molecules and Lewis acids.
→ Ammonia reacts with boron trihalide in a Lewis acid–base reaction with the formation of a dative covalent bond:

$$H_3N\text{:} + BCl_3 \rightarrow H_3N\text{—}BCl_3$$

Sodium tetrahydridoborate(III)

You may meet this compound in organic chemistry questions. Its traditional name is sodium borohydride. $Na^+ BH_4^-$ is not as reactive as its aluminium counterpart so we can use it in aqueous solution to reduce aldehydes and ketones.

Aluminium

→ Aluminium is a highly reactive metal protected by a tough layer of the oxide preventing reaction with the atmosphere.
→ The metal is extracted from *bauxite* ore by electrolysis.
→ Aluminium combines with oxygen, halogens, nitrogen and sulphur at high temperature or when the oxide layer is removed.
→ The oxidation number of aluminium in its compounds is +3.

Aluminium compounds

The high charge and small size of the cation make Al^{3+} strongly polarizing so that many aluminium compounds are covalent.

Aluminium chloride

When dry chlorine is passed over heated aluminium, anhydrous aluminium chloride forms as a covalent sublimate:

$$2Al(s) + 3Cl_2(g) \rightarrow Al_2Cl_6(s)$$

Measurements of M_r in non-aqueous solvents and in the vapour state show that aluminium chloride forms a dimer whose structure can be explained in terms of the electron deficiency of the monomer, $AlCl_3$.

Checkpoint 1

Write the ground state electronic configurations of a B atom and an Al atom.

Grade booster

If you want the top grade make sure you relate boron's atypical properties to the almost unshielded nuclear charge of its small atom.

Grade booster

This reaction of ammonia with boron trichloride is often used to test a top grade student's understanding of bonding and shapes of molecules.

Checkpoint 2

Use the VSEPR theory to predict the shape of
(a) NH_3
(b) $H_3N\text{—}BCl_3$
(c) BH_4^-.

The jargon

Sublime means to change state from solid to vapour (gas) without melting. *Sublimate* is the solid obtained when a vapour (gas) condenses without forming a liquid.

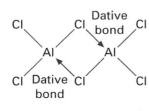

Each Al atom can accept a lone pair of electrons from a Cl atom in the other $AlCl_3$ molecule

Links

See page 14: bonding.

Aluminium oxide

Bauxite, $Al_2O_3 \cdot 2H_2O$, is the mineral from which the metal is extracted by electrolysis. Anhydrous aluminium oxide occurs naturally as corundum and emery. These two very hard minerals are used industrially as abrasives. Ruby, sapphire and topaz are corundum containing chromium, titanium and iron respectively.

→ Aluminium oxide is amphoteric reacting with acids to form $Al(H_2O)_6^{3+}(aq)$ and alkalis to form $Al(OH)_4^-(aq)$.

Aluminium oxide is a refractory material used in furnace linings and in the nose cones of rockets and space vehicles. As a fine powder, it can be used in column **chromatography** and serve as a catalyst in the dehydration of ethanol to ethene.

Grade booster

The bonding in the aluminium chloride dimer is frequently asked. A common error is for students to draw the arrows, representing the dative covalent bonds, the wrong way round. Don't make this mistake if you want the top grade.

Watch out!

You may use washing soda to remove tea stains from cups but *never* from aluminium teapots!

Aluminium hydroxide

When you add NaOH(aq) to aluminium salt solutions you can expect a white gelatinous precipitate to form and then disappear. A simple explanation is that aluminium hydroxide precipitates and then redissolves in excess alkali to form sodium aluminate solution. For a more elaborate interpretation you should refer to

1. the polarizing effect of Al^{3+} ion on H_2O molecules making
2. the aqueous cation act as a Brønsted–Lowry acid $[Al(H_2O)_6]^{3+} \rightleftharpoons [Al(H_2O)_5(OH)]^{2+} + H^+$ and
3. the shift in the equilibrium positions as the addition of OH^- increases the pH causing successive ionizations to produce
4. the white gelatinous precipitate $[Al(H_2O)_3(OH)_3](s)$ and then
5. the colourless solution containing $[Al(OH)_6]^{3-}(aq)$

The jargon

Refractory means able to withstand very high temperatures (>2000°C) without melting or decomposing. *Gelatinous precipitate* is a jelly-like precipitate that settles slowly if at all. *Granular precipitate* is a powder-like precipitate that settles quickly.

Checkpoint 3

State and explain what is formed when you mix sodium carbonate, $Na_2CO_3(aq)$, with aqueous aluminium sulphate to produce a white precipitate and a colourless gas.

Aluminium sulphate and alums

→ *Excess* dilute sulphuric acid dissolves Al(s) and $Al_2O_3(s)$ only very slowly but freshly precipitated $[Al(H_2O)_3(OH)_3](s)$ very quickly to give a solution from which $Al_2(SO_4)_3 \cdot 18H_2O(s)$ will crystallize.

→ *Alum* is potassium aluminium sulphate, $KAl(SO_4)_2 \cdot 12H_2O$, a *double salt* that crystallizes from an aqueous solution containing equal amounts of potassium sulphate and aluminium sulphate.

→ *Alums* are isomorphous double sulphates having the formula $M^IM^{III}(SO_4)_2 \cdot 12H_2O$ where M^I is a group I ion or the ammonium ion and M^{III} usually a transition element ion or the aluminium ion.

The jargon

Isomorphous means having the same shape or angles and capable of overgrowth (one crystal can grow on top of another).

Exam practice (6 minutes)   answers: page 135

Compare and explain the effect of adding aqueous ammonia to separate aqueous solutions of aluminium sulphate and zinc sulphate.

Group IV: elements and oxides

These elements show more strikingly than any other p-block group the typical change in character from non-metal to metal with increasing atomic number down the group.

The elements

→ All five can have an oxidation number of +4 in covalent compounds; they do not form a simple ion with a 4+ charge.
→ Stability of the +2 oxidation state increases down the group with tin and lead able to form $Sn^{2+}(aq)$ and $Pb^{2+}(aq)$.

	Ground state electronic structure	Density /g cm^{-3}	M.p. /°C	B.p. /°C	At. radius /nm
C	$1s^22s^22p^2$	2.26(g) 3.51(d)	3730 sublimes		0.077
Si	$[Ne]3s^23p^2$	2.33	1410	2360	0.117
Ge	$[Ar]3d^{10}4s^24p^2$	5.32	937	2830	0.122
Sn	$[Kr]4d^{10}5s^25p^2$	7.3	232	2270	0.140
Pb	$[Xe]4f^{14}5d^{10}6s^26p^2$	11.4	327	1744	0.154

Carbon
Carbon is the basis of organic life-forms and is the largest group of compounds studied by any one group of chemists. Carbon exists in at least three well-defined allotropic forms. Carbon (graphite) fibres are made from acrylic fibre by heating first in air (to 300 °C) and then in an inert gas (to 1500 °C).

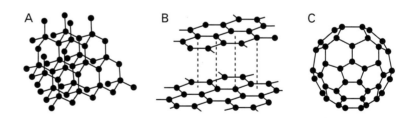

Silicon
Silicon is a shiny, blue-grey solid with a diamond structure but unlike diamond, silicon is a semiconductor used mainly to produce microprocessor chips.

Tin
Tin has two common allotropes – grey tin (diamond structure) and white tin (metallic structure). The metal is used mainly for plating steel but it is also a constituent of bronze – an alloy employed for thousands of years.

Lead
Lead is metallic and used as a roofing material, in shields against radioactive emissions, ionizing radiation and X-rays, in lead/acid car batteries and in solders when alloyed with white tin and other metals.

Checkpoint 1

List three properties of metals which distinguish them from non-metals.

The jargon

Inert pair effect is the tendency for two valence electrons to form a lone pair and not take part in bonding as, for example, in tin(II) and lead(II) ions.
(g) = graphite
(d) = diamond
at. radius = covalent or metallic radius of an atom

Checkpoint 2

(a) What is the name of the allotropes represented by structures A, B, C.
(b) What is a carbon nanotube?

Grade booster

You are not expected to know the chemistry of germanium but be ready to apply your knowledge and understanding of patterns and trends in group IV chemistry if you want to gain the top grade.

The jargon

Tin plating is coating steel with tin.
Galvanizing is coating steel with zinc.

Checkpoint 3

(a) Explain why when scratched and exposed to damp air a tin can rusts but a galvanized plate does not.
(b) Suggest why the demand for and therefore the production of lead is decreasing.

The oxides

Carbon dioxide and carbon monoxide

➜ Carbon dioxide is acidic and is vital in establishing aqueous equilibria that buffer the blood at a constant pH:

$$CO_2(g) + H_2O(l) \rightleftharpoons H_2CO_3(aq) \rightleftharpoons H^+(aq) + HCO_3^-(aq)$$
$$HCO_3^-(aq) \rightleftharpoons CO_3^{2-}(aq) + H^+(aq)$$

➜ Carbon monoxide is neutral, almost insoluble in water and very toxic, displacing oxygen to form carboxyhaemoglobin in blood.

➜ Carbon monoxide will reduce metal oxides and burn in air:

$$CO(g) + PbO(s) \rightarrow CO_2(g) + Pb(s)$$

➜ O═C═O is non-polar and :C≡O: is slightly polar with the oxygen atom being the *positive* end.

:CO and :NO are isoelectronic. The lone pair of electrons (on C and N) makes the molecules powerful ligands capable of forming complexes. For example, the reaction $Ni(s) + 4CO(g) \rightleftharpoons Ni(CO)_4(l)$ is the basis of the Mond process for extracting nickel from its ores or recovering the metal from scrap.

➜ Carbon dioxide is readily identified by the limewater test:

$$CO_2(g) + Ca(OH)_2(aq) \rightarrow CaCO_3(s) + H_2O(l)$$

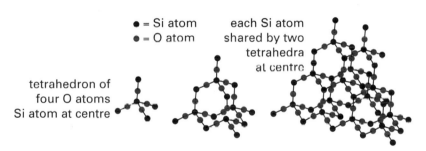

● = Si atom
● = O atom

each Si atom shared by two tetrahedra at centre

tetrahedron of four O atoms Si atom at centre

Silicon dioxide

➜ Silicon dioxide occurs in different forms as a covalent giant molecular structure with the empirical formula SiO_2.

➜ SiO_2 is insoluble in water but regarded as an acidic oxide forming silicates when fused with alkalis, e.g. in the blast furnace or in glass manufacture $CaO + SiO_2 \rightarrow CaSiO_3$ (calcium silicate – slag).

Tin and lead oxides

➜ Tin(II) oxide, SnO, is amphoteric and a powerful reducing agent which is spontaneously oxidized by air.

➜ Lead(II) oxide, PbO, is amphoteric, exists in yellow or red forms and can be prepared by heating the metal in air.

➜ Tin(IV) oxide, SnO_2, is stable and occurs naturally as the ore *cassiterite* from which tin is extracted.

➜ Lead(IV) oxide, PbO_2, is insoluble in water, does not react with dilute acids but forms plumbate(IV) ion, $Pb(OH)_6^{2-}$, with alkalis.

➜ Lead(IV) oxide reacts as an oxidizing agent with concentrated hydrochloric acid: $PbO_2 + 4HCl \rightarrow PbCl_2 + Cl_2 + 2H_2O$.

Don't forget!

You should know the role of carbon dioxide in global warming and that the gas is produced by burning fossil fuels.

Links

See page 150: alkanes.

The jargon

Isoelectronic structures have the same number of electrons arranged in the same way, e.g. CH_4 and NH_4^+.

Watch out!

Carbon monoxide poisons us by binding to the iron(II) in haemoglobin and preventing the blood from transporting oxygen around the body.

Checkpoint 4

How do we account for the use of SiO_2 in (a) furnace linings, (b) abrasives?

The jargon

Hygroscopic means able to absorb moisture from the air.
Deliquescent means able to absorb enough moisture from the air to become an aqueous solution.

Watch out!

Top grade students do not confuse silicon, silica and silicone.

Exam practice (6 minutes)   answers: page 136

(a) State and explain what would be observed when excess carbon dioxide is passed into limewater. (b) Explain the structure and shape of the carbonate ion.

Group IV: chlorides and hydrides

Checkpoint 1

Give the systematic name for
(a) $Pb(NO_3)_2$, (b) $Pb(OH)_4^{2-}$.

The jargon

Red lead is Pb_3O_4 or $PbO_2 \cdot 2PbO$ and shows the oxidizing properties of PbO_2 and the amphoteric character of PbO.

Grade booster

If you want the top grade, you must be able to compare and contrast the group IV oxides. Learn examples of their acidic, amphoteric, basic and oxidizing properties.

Watch out!

Use $HNO_3(aq) + PbO(s)$ for this base dissolving in acid (because $PbCl_2$ and $PbSO_4$ are not very soluble salts).

The jargon

Tetrachloromethane is the systematic name regarding CCl_4 as a substitution product of methane. The traditional name carbon tetrachloride regards CCl_4 as a compound of carbon and chlorine.

Checkpoint 2

(a) Suggest why CCl_4 cannot be prepared by direct reaction of the elements.
(b) Why might aqueous iodine turn colourless and CCl_4 become violet and sink after the liquids are shaken together in a test tube?

Checkpoint 3

Calculate the heat of hydrolysis of $SiCl_4$ to $SiO_2 \cdot 2H_2O$ given their heats of formation to be −620 and −855 kJ mol⁻¹ respectively.

The tetrachlorides of group IV are all covalently bonded and their behaviour is dependent on this. But why is it that CCl_4 has no reaction with water whilst the reaction of $SiCl_4$ and water can be quite dramatic? The ways in which Pb^{2+} can be identified in analysis can also be seen here.

Chlorides

→ Only tin and lead form a chloride in which the element has an oxidation number of +2 and aqueous cation, $Sn^{2+}(aq)$ and $Pb^{2+}(aq)$.
→ All five elements form a covalent tetrachloride whose thermal stability and resistance to hydrolysis decreases down the group.
→ All except CCl_4 hydrolyze to the oxide XO_2 and HCl.

Tetrachloromethane

→ CCl_4 *cannot* be prepared by direct combination of the elements even though the reaction is exothermic and energetically feasible:

$$C(graphite) + 2Cl_2(g) \rightarrow CCl_4(l); \Delta H^{\ominus}_{298} = -129.6 \text{ kJ mol}^{-1}$$

→ Carbon tetrachloride is a dense, colourless liquid used widely as an industrial solvent even though it is carcinogenic and toxic.
→ CCl_4 is *not* attacked even by hot water in spite of the hydrolysis theoretically being exothermic and energetically feasible.

Calculating the standard enthalpy change for the hydrolysis of CCl_4:

$$CCl_4(l) + 2H_2O(l) \rightarrow CO_2(g) + 4HCl(g)$$
$$\Delta H^{\ominus}_{f,298} \quad -130 \quad\quad 2 \times (-286) \quad -394 \quad 4 \times (-92)$$

$\Delta H^{\ominus}_{298}$ is $[(-394) + 4 \times (-92)] - [(-130) + 2 \times (-286)] = -60 \text{ kJ mol}^{-1}$.

→ Hydrolysis of CCl_4 is kinetically hindered because the C-atom *cannot* extend its valence electron shell beyond eight.

Silicon tetrachloride

→ $SiCl_4$ is a colourless liquid which fumes in moist air and hydrolyzes violently on contact with water:

$$SiCl_4(l) + 4H_2O(l) \rightarrow SiO_2 \cdot 2H_2O(s) + 4HCl(g)$$

→ Hydrolysis of $SiCl_4$ is *not* kinetically hindered because the Si-atom *can* extend its valence shell beyond the octet of (eight) electrons.

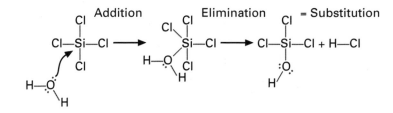

→ Energy released by formation of the strong $H_2O-SiCl_4$ bond provides the energy absorbed by breaking the weak $Si-Cl$ bond.

→ For CCl_4 a strong $C-Cl$ must break before a $C-O$ can form, so the activation energy is very high and the reaction extremely slow.

The products of the hydrolysis would be attacked until all the Cl-atoms have been replaced (substituted) by OH groups.
Elimination of water from $Si(OH)_4$ could produce SiO_2.

Examiner's secrets

Candidates are often asked to explain why $SiCl_4$ is hydrolyzed by water but CCl_4 is not. Remember the Li–F valence shell maximum is 8 electrons, *no 3d*, but the Na–Cl valence shell can extend >8.

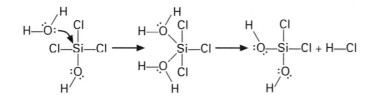

Watch out!

The value of ΔG for the hydrolysis of CCl_4 as shown by the equation above on page 102, is negative.

Tin(IV) chloride and lead(IV) chloride
→ These are covalent liquids hydrolyzed by water and moist air.
→ Both will dissolve in concentrated hydrochloric acid to form complex hexachloroanions $[SnCl_6]^{2-}$ and $[PbCl_6]^{2-}$.

Tin(II) chloride and lead(II) chloride
→ Both chlorides are regarded as ionic with some covalent character.
→ $SnCl_2(aq)$ is dissolved in HCl(aq) to prevent hydrolysis and kept in contact with metallic tin to prevent oxidation to the tin(IV) state.
→ $PbCl_2(aq)$ contains $Pb^{2+}(aq)$ but $PbCl_2(s)$ precipitates when a hot solution cools because lead(II) chloride is a sparingly soluble salt.
→ Tin(II) chloride is a strong reducing agent:

Checkpoint 4

What is the systematic name for the $[SnCl_6]^{2-}$ and $[PbCl_6]^{2-}$ anions?

$$SnCl_2 + 2HgCl_2 \rightarrow SnCl_4 + Hg_2Cl_2 \text{ (white precipitate)}$$
$$Hg_2Cl_2 + SnCl_2 \rightarrow SnCl_4 + 2Hg \text{ (grey turning to silver liquid)}$$

Hydrides

All five elements form covalent tetrahydrides but the stability decreases down the group with PbH_4 formed only in traces.
Si forms silanes, Si_nH_{2n+2}, but all hydrolyze even in moist air.

Don't forget!

The first member of the group will show some atypical behaviour and the trend down the group with increasing atomic number is increasing metallic character and decreasing stability of the +4 oxidation state.

Exam practice (10 minutes)   answers: page 136

(a) Give equations and outline how, starting from the element, you would make a sample of (i) silicon tetrachloride, and (ii) lead(II) chloride.

(b) Explain what happens on adding (i) silicon tetrachloride to aqueous sodium hydroxide, and (ii) concentrated hydrochloric acid to lead(II) chloride.

Group V: elements and oxides

These elements typically change character from non-metal to metal with increasing atomic number down the group.

The elements

→ All five can have oxidation numbers of +5 and +3 in their compounds but the stability of the +5 oxidation state decreases with increasing atomic number down the group from N to Bi.
→ Nitrogen is the only element to triple bond with itself $N\equiv N$.
→ White phosphorus and one form of both antimony and arsenic can exist as tetraatomic molecules in which each atom is attached to the other three by single covalent bonds.

	Ground state electronic structure	Density/g cm^{-3}	M.p./°C	B.p./°C
N	$1s^2 2s^2 2p^3$	0.81 at 77 K	−210	
P	$[Ne]3s^2 3p^3$	1.82 white	44.2 white	
		2.34 red	590 red	
As	$[Ar]3d^{10}4s^2 4p^3$	5.72 grey		613 sublimes
Sb	$[Kr]4d^{10}5s^2 5p^3$	6.62	630	1 380
Bi	$[Xe]4f^{14}5d^{10}6s^2 6p^3$	9.80	271	1 560

Nitrogen

→ Nitrogen is about 78% by volume of our atmosphere from which it can be extracted by liquefaction and fractional distillation.
→ Nitrogen exists as very stable diatomic gas molecules and shows some properties atypical of group V.
→ Nitrogen will combine with lithium (and other very electropositive metals) to form the nitride ion, N^{3-}.
→ Nitrogen will combine at high temperatures with non-metals such as boron, hydrogen and oxygen to form BN, NH_3 and oxides.

$$2B(s) + N_2(g) \rightarrow 2BN(s) \text{ (structure similar to graphite)}$$
$$3H_2(g) + N_2(g) \rightleftharpoons 2NH_3(g) \text{ (nitrogen fixation by the Haber process)}$$
$$N_2(g) + O_2(g) \rightarrow 2NO(g) \text{ (lightning or spark discharging in air)}$$

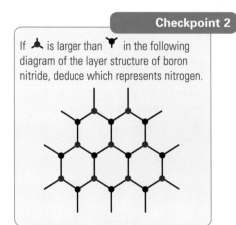
Phosphorus

→ Phosphorus has two allotropes.
→ White phosphorus, P_4, is very reactive and stored under water.
→ Red phosphorus is a less reactive polymeric form of the element.
→ Phosphorus combines directly with oxygen and chlorine forming oxides and chlorides in which its oxidation state can be +3 and +5.

$$4P(s) + 3O_2(g) \rightarrow P_4O_6(s); \quad 4P(s) + 5O_2(g) \rightarrow P_4O_{10}(s) \text{ (excess } O_2)$$
$$2P(s) + 3Cl_2(g) \rightarrow 2PCl_3(l); \quad 2P(s) + 5Cl_2(g) \rightarrow [PCl_4]^+[PCl_6]^-(s)$$

Arsenic, antimony and bismuth

→ The chemistry of these three elements is not important for A-level except for illustrating the group trend from non-metal to metal.

Oxides

Oxides of nitrogen

→ The oxidation state of nitrogen varies from +1 to +5 in its oxides N_2O, NO, N_2O_3 (very unstable), NO_2 ($\rightleftharpoons N_2O_4$), N_2O_5 (unstable).

The oxides have *delocalized structures* that cannot be represented using only '·', '—' and '→' for an electron, a shared electron pair and a donated pair of electrons.

$$:N\equiv N \rightarrow \ddot{O}: \qquad :\ddot{N}=\ddot{O}:$$

The jargon

Traditional names: *nitrous oxide (laughing gas)* = N_2O; *nitric oxide* = NO.

Links

See page 17: delocalized bonding.

→ Dinitrogen oxide (nitrogen(I) oxide) is a colourless gas readily decomposed by a red-hot wooden splint: $2N_2O \rightarrow 2N_2 + O_2$.

→ Nitrogen (mon)oxide, NO, is an odd-electron molecule that reacts spontaneously with oxygen to form NO_2 (brown gas).

→ NO_2 dimerizes to dinitrogen tetraoxide, N_2O_4 (colourless gas).

→ On warming, dinitrogen tetraoxide (colourless) decomposes reversibly into nitrogen dioxide (brown): $N_2O_4(g) \rightleftharpoons 2NO_2(g)$.

→ At 150 °C the gas contains almost 100% of NO_2 molecules and at higher temperatures the nitrogen dioxide decomposes into nitrogen oxide and oxygen: $2NO_2(g) \rightleftharpoons 2NO(g) + O_2(g)$.

→ Dinitrogen pentoxide, $NO_2^+NO_3^-(s)$, decomposes above its melting point in a first order reaction: $N_2O_5(s) \rightarrow 2N_2O_4(g) + O_2(g)$.

Checkpoint 3

Explain why a glowing splint rekindles when put into a test tube of N_2O.

Grade booster

For the top grade make sure you understand how to apply (a) the law of chemical equilibrium to the decomposition of N_2O_4 and (b) first order kinetics to the decomposition of N_2O_5.

Oxides of phosphorus

→ Two solid oxides are produced by burning phosphorus in (limited or excess) air: $4P + 3O_2 \rightarrow P_4O_6$ and $4P + 5O_2 \rightarrow P_4O_{10}$.

→ In the gas phase, the structure of the oxide molecules is based on a tetrahedron of phosphorus atoms.

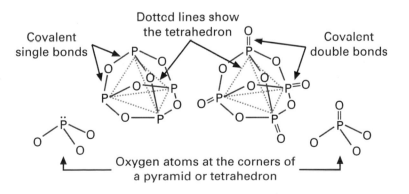

Oxygen atoms at the corners of a pyramid or tetrahedron

Don't forget!

Both nitrogen and phosphorus are important elements in plant growth, and biochemistry generally.
Balanced fertilizers, sometimes called NPK fertilizers, contain the essential elements nitrogen, phosphorus and potassium. The production of fertilisers is an important contribution that chemistry makes to society.

→ Phosphorus(V) oxide reacts violently with water and is used in organic chemistry as a powerful dehydrating agent.

Exam practice (10 minutes)   answers: page 136

(a) Give an equation and explain why (i) nitrogen oxide turns brown in air, and (ii) dinitrogen oxide rekindles a glowing splint.

(b) Give the name and formula of the sodium salt formed when excess sodium hydroxide reacts with (i) N_2O_3, (ii) N_2O_5, (iii) P_4O_6, (iv) P_4O_{10}.

Group V: oxoacids and hydrides

For A-level chemistry, the three most important oxoacids are nitric(V) acid, nitric(III) acid and phosphoric(V) acid.

The jargon

Nitrous acid is the traditional name for nitric(III) acid.
In situ means in the place (reaction mixture) where and when needed.
Nitric acid is an acceptable name for nitric(V) acid.

Links

See page 160: diazotization.

Checkpoint 1

Write an equation for the reaction of $N_2O_4(g)$ with NaOH(aq).

Checkpoint 2

(a) What is the pH of aqueous nitric acid of concentration 0.02 mol dm⁻³?
(b) How would you prepare a sample of sodium nitrate in the laboratory from nitric acid?

Grade booster

For the top grade make sure you can write the balanced equations for the reduction of dilute and concentrated aqueous nitric acid by copper.

Checkpoint 3

What is the conjugate base of H_3PO_4?

The jargon

Phosphoric acid = orthophosphoric acid = tetraoxophosphoric(V) acid.

Nitrous acid

→ Nitrous acid, HNO_2, disproportionates into nitric acid and nitrogen monoxide and is also rapidly oxidized by air to nitric acid:

$$3HNO_2(aq) \rightarrow HNO_3(aq) + H_2O(l) + 2NO(g)$$
oxidation nos.   +3   +5   +2

→ We usually use ice-cold aqueous sodium nitrite and HCl(aq) to prepare HNO_2(aq) *in situ* for diazotizations in organic chemistry:

$$NaNO_2(aq) + HCl(aq) \rightarrow HNO_2(aq) + NaCl(aq)$$

→ Sodium (or potassium) nitrite is produced when sodium (or potassium) nitrate is heated: $2NaNO_3(s) \rightarrow 2NaNO_2(s) + O_2(g)$.

An aqueous mixture of nitrite and nitrate is produced when nitrogen dioxide (and dinitrogen tetraoxide) reacts with water or alkali.

Nitric acid

→ Nitric acid, HNO_3, can be collected as a yellow distillate when a mixture of sodium nitrate and concentrated sulphuric acid is heated:

$$NaNO_3(s) + H_2SO_4(l) \rightarrow NaHSO_4(s) + HNO_3(l)$$

→ The acid is coloured by nitrogen dioxide formed by thermal decomposition of some nitric acid vapour:

$$2HNO_3(g) \rightarrow H_2O(l) + 2NO_2(g) + \tfrac{1}{2}O_2(g)$$

→ Nitric acid is manufactured on the industrial scale by the catalytic oxidation of ammonia.

In dilute aqueous solution, nitric acid behaves as a typical strong monobasic acid when it reacts with alkalis and carbonates but not when it reacts with metals above hydrogen in the electrochemical series. For example, magnesium reduces NO_3^-(aq) to NH_4^+(aq) and not H^+(aq) to H_2(g):

→ Aqueous nitric acid does *not* react with metals to form hydrogen.
→ Nitric acid is a strong oxidant readily reduced by copper to NO_2 or NO depending upon the acid concentration:

$$Cu + 4HNO_3(conc.) \rightarrow Cu(NO_3)_2 + 2H_2O + 2NO_2$$
$$3Cu + 8HNO_3(dil.) \rightarrow 3Cu(NO_3)_2 + 4H_2O + 2NO$$

All nitrates (salts of nitric acid) are soluble in water and decompose on heating to give off oxygen and (except for the alkali metal nitrates) poisonous brown fumes of nitrogen dioxide:

$$2Cu(NO_3)_2 \rightarrow 2CuO + 4NO_2 + 2O_2$$

Oxoacids of phosphorus

Phosphorus has a larger range of oxoacids than nitrogen in spite of having a smaller range of oxides than nitrogen.

→ P_4O_6 is the anhydride of phosphonic acid, H_3PO_3, and P_4O_{10} is the anhydride of phosphoric acid, H_3PO_4.

Chlorides

Nitrogen forms only the trichloride (not important for A-level) but phosphorus forms the trichloride and the pentachloride by direct reaction with the appropriate amounts of chlorine.

Phosphorus chlorides

→ $PCl_3(l)$ and $PCl_5(s)$ react vigorously with O–H (in water, alcohols, etc.) and N–H (in NH_3 and amines) to give steamy fumes of HCl:

$$2PCl_3 + 3ROH \rightarrow 3RCl + H_3PO_3 + 3HCl$$

→ PCl_3 and gaseous PCl_5 are covalent but *solid* pentachloride has an ionic structure.

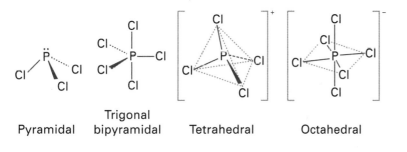

| Pyramidal | Trigonal bipyramidal | Tetrahedral | Octahedral |

Hydrides

Both elements form two covalent hydrides but only ammonia, NH_3, and hydrazine, N_2H_4, are important for A-level.

Ammonia

→ Ammonia is manufactured from N_2 and H_2 by the Haber process.
→ Ammonia gas (b.p. –33 °C) liquefies easily, dissolves well in water forming a weak alkali and forms ammonium salts with acids:

$$NH_3(aq) + H_2O(l) \rightleftharpoons NH_4^+(aq) + OH^-(aq)$$

→ We use the formation of a white smoke with hydrogen chloride as a test for either gas:

$$NH_3(g) + HCl(g) \rightleftharpoons NH_4Cl(s)$$

→ Ammonia is given off when ammonium salts are heated with s-block alkalis:

$$NH_4^+(aq) + OH^-(aq) \rightarrow NH_3(g) + H_2O(l)$$

→ Ammonium salts readily dissolve in water and dissociate or decompose on heating:

$$NH_4NO_3(s) \rightarrow N_2O(g) + 2H_2O(l)$$
$$NH_4NO_2(aq) \rightarrow N_2(g) + 2H_2O(l)$$

→ Ammonia is an important industrial chemical used mainly for the production of fertilizers, nitric acid (by oxidation) and nylon.

Exam practice (6 minutes)   answers: page 137

(a) Write a balanced equation and state the oxidation number of nitrogen in each product when dinitrogen tetraoxide reacts with aqueous sodium hydroxide.

(b) Calculate (i) the mass of water needed to convert 0.1 mol PCl_3 into H_3PO_3 and (ii) the maximum theoretical volume of HCl(g) produced at room temperature. [molar gas volume = 24 dm³ at room temperature]

Watch out!

NCl_3 is a dangerously explosive yellow oil which maimed Sir Humphrey Davy!

Checkpoint 4

Write an equation for the hydrolysis of phosphorus pentachloride.

Grade booster

We should expect top grade students to be able to draw clear diagrams of the structures of these phosphorus chlorides.

Links

See page 119: the Haber process.

Checkpoint 5

Suggest why, when some $NH_4Cl(s)$ in the bottom of a test tube is gently warmed, (a) a white deposit appears higher up inside the tube and (b) moist red litmus paper held near but above the mouth of the test tube turns blue.

Watch out!

$NH_4NO_2(aq)$, made by mixing $NH_4Cl(aq)$ and $NaNO_2(aq)$, could decompose explosively when heated!

Group VI:
oxygen and sulphur

Only these first two non-metals are important for A-level. The trends with increasing atomic number down the group follow similar patterns to those of other p-block groups.

The elements

→ Both elements exist as simple covalent molecules, O_2 and S_8.
→ Both elements have allotropes: O_2 and O_3 (ozone) in the gas phase, rhombic and monoclinic sulphur in the solid (crystalline) state.
→ Both elements are components of rocks and minerals in the earth's crust and exist uncombined (oxygen in air, sulphur in the ground).
→ Oxygen is extracted (blue liquid, b.p. –183 °C) from the atmosphere by fractional distillation of liquid air while sulphur is extracted from the ground in the Frasch process.
→ Both elements combine with metals to form anions (O^{2-} and S^{2-}) and with non-metals to form covalent molecules (CO_2 and CS_2).

Oxygen

→ Oxygen is formed by thermal decomposition of metal nitrates and by electrolytic decomposition of water using an inert anode and a suitable electrolyte (e.g. H_2SO_4) to improve conductivity:

$$4OH^-(aq) \rightarrow 2H_2O(l) + O_2(g) + 4e^- \text{ (anodic oxidation –2 to 0)}$$

→ Oxygen is produced by green plants during the photosynthesis of carbohydrates from atmospheric carbon dioxide and water:

$$6CO_2(g) + 6H_2O(l) \rightarrow C_6H_{12}O_6(s) + 6O_2(g)$$

Uses of oxygen

→ steel production – burns off unwanted carbon in pig-iron
→ metal cutting and welding – burns $C_2H_2(g)$ in oxyacetylene torch
→ medical uses and as ozone for air and water sterilization

Trioxygen, O_3 occurs in the stratosphere in the ozone layer which absorbs and protects the earth from harmful UV radiation.

→ Ozone is formed when an electric discharge passes through air (or oxygen): $3O_2(g) \rightarrow 2O_3(g) \rightarrow 3O_2(g)$.

This ozonized air (or oxygen) may contain up to 10% ozone. The O_3 is *not* in an equilibrium with O_2 but its decomposition back to oxygen is very slow at room temperature.

Sulphur

Rhombic (the more stable) and monoclinic sulphur crystals consist of closely packed S_8 covalent molecules (buckled rings).

On heating in a test tube, sulphur melts to a yellow liquid, darkens, and suddenly becomes unexpectedly extremely viscous. On heating further, the orange-red jelly becomes less viscous and turns almost black as the sulphur nears its boiling point. At the mouth of the tube a blue flame may appear.

→ Most metals including silver form sulphides many of which occur naturally as important ores mined for the extraction of the metal.

→ Sulphur burns in oxygen to sulphur dioxide.

Oxides and oxoacids

Binary compounds with oxygen exist for most elements. We classify these in several ways. At A-level you can classify them by *structure* (ionic, covalent, giant covalent, peroxide, superoxide) and by *acid–base behaviour* (basic, acidic, amphoteric and neutral).

Sulphur dioxide and trioxide

→ Sulphur dioxide liquefies easily, dissolves well in water forming a weak acid and forms sulphites with alkalis:

$$SO_2(g) + H_2O(l) \rightarrow H_2SO_3(aq) \text{ (sulphurous acid)}$$

If you add dilute hydrochloric acid to a sulphite, you will detect $SO_2(g)$ given off. Aqueous sulphur dioxide and sulphites are reducing agents that turn $Cr_2O_7^{2-}(aq)/H^+(aq)$ from orange to green.

→ Sulphur trioxide has a trigonal planar molecule in the gaseous state and forms deliquescent crystalline needles and a polymer in the solid state.

→ Sulphur trioxide is formed from SO_2 and O_2 in the Contact process for the manufacture of sulphuric acid, H_2SO_4.

Sulphurous and sulphuric acid

→ Both acids are dibasic, forming normal and acid salts with alkalis.

→ $H_2SO_3(aq)$ is a weak acid and $H_2SO_4(aq)$ is a strong acid:

$$H_2SO_3(aq) \rightleftharpoons H^+(aq) + HSO_3^-(aq) \rightleftharpoons H^+(aq) + SO_3^{2-}(aq)$$
$$H_2SO_4(aq) \rightarrow H^+(aq) + HSO_4^-(aq) \rightleftharpoons H^+(aq) + SO_4^{2-}(aq)$$

→ Sulphurous acid exists only in aqueous solution but (anhydrous) sulphuric acid is a viscous colourless liquid (b.p. 280 °C).

To make dilute aqueous sulphuric acid we add drops of conc. H_2SO_4 to a large volume of water stirred vigorously to dissipate the large amount of heat produced. Conc. H_2SO_4 can be reduced to SO_2 by metals and non-metals, e.g. $Zn(s) + 2H_2SO_4(l) \rightarrow ZnSO_4(aq) + SO_2(g) + 2H_2O(l)$ and $C(s) + 2H_2SO_4(l) \rightarrow CO_2(g) + 2SO_2(g) + 2H_2O(l)$. Conc. H_2SO_4 can be used as a dehydrating agent:

$(COOH)_2(s) \rightarrow CO(g) + CO_2(g) + H_2O(l)$.

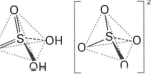

→ The sulphuric acid molecule and the sulphate anion have tetrahedral structures, the SO_4^{2-} being delocalized with four identical bonds between S—O and S=O equal in length.

Action point

Make a table to show the two classes (ionic or covalent) of binary oxygen compounds and include Al_2O_3, CsO_2, CaO, CO_2, CO, K_2O_2, NO_2, SiO_2, Na_2O as examples.

The jargon

Sulphurous acid = sulphuric(IV) acid
Sulphuric acid = sulphuric(VI) acid
Sulphite = sulphate(IV)
Sulphate = sulphate(VI)

Watch out!

Sulphur dioxide may be used to sterilize food and wine but it is toxic. Fortunately its characteristic sharp smell will warn you of danger.

Links

See page 120: the Contact process.

The jargon

Normal salt = all protons donated:
e.g. $H_3PO_4 \rightarrow Na_3PO_4$
Acid salt = *not* all protons donated:
e.g. $H_3PO_4 \rightarrow Na_2HPO_4$ and NaH_2PO_4

Watch out!

pH of acid salt solutions need not be <7! $Na_2HPO_4(aq)$ is alkaline: pH > 7. Only the first ionization of $H_2SO_4(aq)$ is complete. The second is much weaker with pK_a = 1.99. Top grade candidates would understand and remember this.

Exam practice (10 minutes)
answers: page 137

(a) Suggest an explanation for (i) the changes observed when sulphur is heated and (ii) what happens when the hot liquid sulphur is poured quickly into cold water.

(b) Write equations for the reaction of sulphur with (i) magnesium and (ii) chlorine

(c) Suggest why the oxidation states of sulphur in Na_2S and SF_6 are different.

Group VI: water and hydrogen peroxide

Both oxygen and sulphur form hydrides, H_2O and H_2S. We die without water. We can be poisoned by hydrogen sulphide (TLV 20). Oxygen forms a peroxide, H_2O_2. Sulphur can form aqueous polysulphide anions.

Watch out!

$H_2(g)$ reacts with $O_2(g)$ explosively and irreversibly but with boiling sulphur only very slightly and reversibly.

Checkpoint 1

What is the conjugate base of (a) H_3O^+ and (b) OH^-? Suggest why water may be seen as amphoteric.

Water

→ Water is the most abundant liquid on our planet.
→ Water is a poor conductor (but not an insulator) and its self-ionization defines neutrality for aqueous acid–base systems:

$$H_2O(l) + H_2O(l) \rightleftharpoons H_3O^+(aq) + OH^-(aq)$$

Compared to most materials or to covalent compounds of similar molar mass, water has some exceptional properties:

1 high melting point and boiling temperatures
2 high specific heat capacity and heats of fusion and vaporization
3 solid (ice) less dense than the liquid on which it floats
4 density of the liquid has a maximum value at $4\,°C$

You should be able to account for these properties in terms of the structure, shape and bonding of the polar water molecule and the intermolecular forces in the liquid and in the open structure of the solid.

104.5°

hydrogen bonding

Links

See page 19–20: hydrogen bonding.

→ Water is an excellent polar solvent for ionic solids and many polar covalent compounds.

Action point

Find out how the hardness of water (in mg calcium carbonate per litre) would be determined by titration and what range of values would classify water as soft or hard.

Water from limestone districts is 'hard' because it contains $Ca^{2+}(aq)$ which precipitates (soluble sodium) soaps as 'scum' (insoluble calcium soaps): $Ca^{2+}(aq) + 2C_{17}H_{35}COO^-(aq) \rightarrow (C_{17}H_{35}COO^-)_2Ca^{2+}(s)$. Some so-called 'temporary hardness' is removed by boiling if $HCO_3^-(aq)$ is present: $Ca^{2+}(aq) + 2HCO_3^-(aq) \rightarrow CaCO_3(s) + H_2O(l) + CO_2(g)$. The precipitated calcium carbonate is the limescale or kettle fur. $Mg^{2+}(aq)$ and any $Ca^{2+}(aq)$ not removed by boiling is 'permanent hardness'.

→ Synthetic ion-exchange resins (used in dishwashers) soften water by absorbing $Ca^{2+}(aq)$ and replacing it with $Na^+(aq)$.
→ Polyphosphates (included in washing powders) can soften water by forming complexes with the $Ca^{2+}(aq)$.

Checkpoint 2

Suggest why some dishwashers have a container that must be kept filled with granular salt (sodium chloride).

Many crystalline salts are hydrated, e.g. cobalt(II) chloride-6-water and copper(II) sulphate-5-water. Some or all of the water molecules are attached to the cation by dative covalent bonds. The O-atom in the H_2O donates the lone pair for sharing and forming the bond.

Watch out!

We prefer to use sodium sulphate as a drying agent because calcium chloride is deliquescent and may not be so easy to separate from the organic liquid.

→ Water molecules can act as ligands to form complexes.
→ We use $CaCl_2(s)$ and $Na_2SO_4(s)$ as drying agents in organic chemistry because they readily form hydrates $CaCl_2 \cdot 6H_2O(s)$ and $Na_2SO_4 \cdot 10H_2O(s)$ but do not dissolve in organic liquids.
→ Water may act as a nucleophile in the hydrolysis of some inorganic and organic halogen compounds.

Links

See pages 112: inorganic halides; and 155: acyl chlorides.

Water reacts as a reducing agent with fluorine because the halogen is a more powerful oxidant than oxygen itself and displaces it:

$$2F_2(g) + 2H_2O(l) \rightarrow 4HF(aq) + O_2(g)$$

→ Water normally reacts as an oxidant and is reduced to hydrogen by reaction with, for example, metals and methane:

$$2Na(s) + 2H_2O(l) \rightarrow 2NaOH(aq) + H_2(g)$$

$$CH_4(g) + 2H_2O(g) \xrightarrow[\text{nickel catalyst}]{900\,°C\ 30\ atm} CO(g) + 2H_2(g)$$

Hydrogen peroxide

H_2O_2 is formed in a variety of situations including glow discharge through water vapour, oxidation of hydrocarbons, exposing a hydrogen–oxygen mixture to intense UV light and during electrolysis of aqueous sulphuric acid and sulphates. In many of these cases, hydroxyl **free radicals** may be formed and, under suitable conditions, may combine to give the peroxide: $2HO\cdot \rightarrow HO{-}OH$.

→ Pure hydrogen peroxide is a pale blue liquid liable to decompose explosively and exothermically without warning into H_2O and O_2.

You should be able to use the VSEPR theory to predict a shape for the molecule.

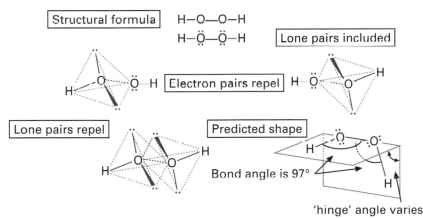

'hinge' angle varies

s-Block peroxides are ionic with O_2^{2-} ions in their structure
We could prepare aqueous hydrogen peroxide in the laboratory from barium peroxide and ice-cold aqueous sulphuric acid because barium sulphate is insoluble: $BaO_2(s) + H_2SO_4(aq) \rightarrow BaSO_4(s) + H_2O_2(aq)$. However, we usually buy aqueous hydrogen peroxide sold as '20 (or 100) volume solution'.

→ Peroxides can provide free radicals to initiate **polymerizations**.

Potassium manganate(VII) oxidizes hydrogen peroxide in excess $H_2SO_4(aq)$ quantitatively:
$$5H_2O_2(aq) + 2MnO_4^-(aq) + 6H^+(aq) \rightarrow 2Mn^{2+}(aq) + 8H_2O(l) + 5O_2(g).$$

Exam practice (4 minutes)   answers: page 137

25.0 cm³ of aqueous hydrogen peroxide in excess acid required 37.5 cm³ of aqueous potassium manganate(VII), of concentration 0.018 7 mol dm⁻³, for complete oxidation. Calculate the concentration of aqueous hydrogen peroxide in mol dm⁻³.

Links

See page 148: free radical substitution.

Watch out!

109.5° = tetrahedral angle but bonded (bp) and non-bonded (np) repulsive forces vary np:np>np:bp>bp:bp so in a in water molecule the H–O–H bond angle is 104.5°.

Grade booster

You won't be asked anything more difficult than predicting the shape of hydrogen peroxide. For the top grade you should be able to explain the shape in terms of the electron-pair and bond-pair repulsions.

The jargon

$O_2^{2-} = (O{-}O)^{2-}$ the *peroxide ion*
$O_2^- = (O{-}O)^-$ the *superoxide ion*
'20 volume solution' means that 1 cm³ of the solution yields 20 cm³ of oxygen at s.t.p. on decomposition.

Links

See page 152: free radical polymerization of ethene and chloroethene.

Watch out!

Aqueous hydrogen peroxide is normally reduced because it is a powerful oxidizing agent (bleaching hair in this way).

Group VII: halogens and hydrogen halides

The halogens are very important at A-level. They are all non-metals and demonstrate very clear trends in their behaviour as you go down the group. Most syllabuses do not include fluorine but check yours just in case.

The elements

→ All can have an oxidation number of –1 in covalent compounds and form a simple anion with a 1– charge.
→ Fluorine is a stronger oxidant than oxygen and forms compounds with all other elements except helium, neon and argon.
→ Unlike fluorine, the other halogens can have positive oxidation numbers and undergo disproportionation reactions.
→ All the halogens are toxic but chlorine is used to sterilize water and iodine has been used as an antiseptic agent.

Formula and state	Colour of element	Electronic configuration	M.p. /°C	B.p. /°C	E(X–X) /kJ mol^{-1}	E(H–X) /kJ mol^{-1}	Np
$F_2(g)$	pale yellow	$1s^2 2s^2 2p^5$	–220	–188	158	562	4.0
$Cl_2(g)$	greenish-yellow	$1s^2 2s^2 2p^6 3s^2 3p^5$	–101	–35	242	431	3.0
$Br_2(l)$	red-brown	$[Ar]3d^{10} 4s^2 4p^5$	–8	59	193	366	2.8
$I_2(s)$	lustrous grey-black	$[Kr]4s^2 4p^6 4d^{10} 5s^2 5p^5$	114	184	151	299	2.5

Fluorine

→ $F_2(g)$ is extremely dangerous and its reactions very exothermic, often forcing other elements into their highest oxidation state.
→ Most metals catch fire in fluorine and water reacts to form a number of products including O_2, O_3 and H_2O_2.

Chlorine

→ $Cl_2(g)$ is denser than air and soluble in water where some of it disproportionates: $Cl_2(aq) + H_2O(l) \rightleftharpoons HCl(aq) + HClO(aq)$.
→ $Cl_2(g)$ oxidizes many metals to their higher oxidation state: $2Fe(s) + 3Cl_2(g) \rightarrow 2FeCl_3(s)$.
→ $Cl_2(g)$ forms compounds with many non-metals (e.g. PCl_3 and PCl_5) but does not react directly with carbon, nitrogen and oxygen.
→ Chlorine can displace bromine and iodine from their compounds and from their aqueous ions in accord with the $E^{\ominus}/V$ values for the following redox processes:

$$Cl_2(aq) + 2e^- \rightleftharpoons 2Cl^-(aq) \quad + 1.36$$
$$Br_2(aq) + 2e^- \rightleftharpoons 2Br^-(aq) \quad + 1.09$$
$$I_2(aq) + 2e^- \rightleftharpoons 2I^-(aq) \quad + 0.54$$
$$F_2(aq) + 2e^- \rightleftharpoons 2F^-(aq) \quad + 2.87$$

Note: Fluorine would never be used in aqueous displacement reactions. Chlorine disproportionates in NaOH(aq) to different oxidation states depending on the alkali concentration and temperature:

1 $Cl_2(g) + 2NaOH(aq) \rightarrow NaCl(aq) + NaClO(aq) + H_2O(l)$
2 $3Cl_2(g) + 6NaOH(aq) \rightarrow 5NaCl(aq) + NaClO_3(aq) + 3H_2O(l)$

Some manufacturing processes produce excess chlorine – a potential environmental health hazard. This chlorine can be absorbed by alkalis to produce for example NaClO (sodium chlorate(I) or sodium hypochlorite), the active ingredient in household bleach. $NaClO_3$ (sodium chlorate(V) or sodium chlorate) is a weed killer and oxidant.

Bromine and iodine

→ Bromine is moderately and iodine only slightly soluble in water.
→ Halogens dissolve in organic solvents, iodine solutions being brown in alcohols and violet in hydrocarbons or halogenoalkanes.
→ Br_2 and I_2 reactions are like those of chlorine but less vigorous.
→ $Br_2(aq)$ is rapidly decolourized by alkenes and $I_2(aq)$ forms a blue-black complex with starch (a very sensitive test for I_2 or starch).

Hydrogen halides

→ All four (volatile) hydrogen halides are released as gases from solid halides by hot (involatile) phosphoric acid:
$Khal(s) + H_3PO_4(l) \rightarrow KH_2PO_4(s) + Hhal(g)$.
→ They dissolve readily in water to give (weak) hydrofluoric acid and strong hydrochloric, hydrobromic and hydroiodic acids.
→ Strong oxidizing agents like $KMnO_4(s)$ or $PbO_2(s)$ release the halogen when warmed with conc. HCl(aq), HBr(aq) and HI(aq):
$PbO_2(s) + 4HCl(conc.aq) \rightarrow PbCl_2(s) + 2H_2O(l) + Cl_2(g)$.
→ We can detect hydrogen halides by the white smoke of ammonium halide formed exothermically on contact with ammonia gas.

Hydrogen fluoride

Hydrogen fluoride is prepared by the action of hot concentrated sulphuric acid on an ionic fluoride: $CaF_2(s) + H_2SO_4(l) \rightarrow CaSO_4(s) + 2HF(g)$. It etches glass.

→ Hydrogen fluoride is highly polar and hydrogen bonded.

High intermolecular forces are the reason why the boiling point (19.5 °C) is higher than that of HCl. Unusually strong hydrogen bonds cause dimers, H_2F_2, and lower the number of protons donated in water, so the aqueous acid is weak ($K_a = 7 \times 10^{-4}$ mol dm^{-3}). Sodium hydrogenfluoride, $NaHF_2(s)$, a salt with HF_2^- ions, exists.

→ Hydrofluoric acid is stored in polythene containers because it attacks and etches glass by reacting with the silicon dioxide:
$SiO_2(s) + 6HF(aq) \rightarrow H_2SiF_6(aq)$.

Grade booster
You should be prepared for halogen chemistry to appear in many different contexts linked, for example, to electrochemistry, organic reaction mechanisms, redox, etc., if you want to get the top grade.

Watch out!
Even bromine water, $Br_2(aq)$, should be used in a fume cupboard.

Links
See page 151: alkenes and halogens.

Checkpoint 2
Write an equation for the reaction of
(a) hydrogen chloride with ammonia
(b) conc. hydrochloric acid with $KMnO_4(s)$

Watch out!
Hydrogen fluoride, whether liquid, vapour or aqueous, is extremely dangerous. The small fluoride ion penetrates deep into tissue to cause very painful burns!

Exam practice (5 minutes)   answers: page 137

(a) Write an equation and give the oxidation number of the chlorine in each product when the gas reacts with (i) cold, dilute NaOH(aq), (ii) hot, concentrated NaOH(aq). (b) Predict the structure and shape of the HF_2^- ion.

Group VII: halides and interhalogen compounds

Now we look at the compounds formed between the halogens and hydrogen and their stability, which you can relate to the increasing bond length from HCl to HI. You will also see the ways in which solid and aqueous halides can be identified.

Links

See page 121: HCl manufacture.

Examiner's secrets

You won't get a mark if you write $2NaCl(s) + H_2SO_4(l) \rightarrow Na_2SO_4(s) + 2HCl(g)$ – the test tube would melt before this reaction could happen!

Checkpoint 1

Deduce which of HBr and HI is the stronger reductant by finding the change in oxidation number of sulphur when each reduces sulphuric acid.

Examiner's secrets

Questions are often set on the relative stabilities of the hydrogen halides. You will get marks for saying that H–Hal bond energy decreases and bond length increases from HCl – HBr – HI.

Checkpoint 2

Write an equation for the reaction of sodium hydrogencarbonate with hydrochloric acid.

Grade booster

Questions are often set on identifying halide in solids and halide ions in aqueous solutions. You should know what to do, what to observe and what equations to write to explain the chemistry behind the observations if you want the top grade.

Hydrogen chloride

Hydrogen chloride is prepared by the action of hot concentrated sulphuric acid on an ionic chloride: $NaCl(s) + H_2SO_4(l) \rightarrow NaHSO_4(s) + HCl(g)$.

→ Hydrogen chloride is a colourless, choking gas that fumes in moist air and dissolves rapidly in water where as a strong acid it ionizes completely: $HCl(g) + H_2O(l) \rightarrow H_3O^+(aq) + Cl^-(aq)$.

Hydrogen bromide and hydrogen iodide

Like hydrogen chloride, HBr(g) and HI(g) are colourless choking gases that fume in moist air and dissolve rapidly in water where as strong acids they ionize completely into hydrogen ions and halide ions.

→ Hydrogen bromide and hydrogen iodide are *not* prepared using hot conc. sulphuric acid because

1   they are reductants and would be oxidized by the acid:

$$2HBr(g) + H_2SO_4(l) \rightarrow 2H_2O(l) + SO_2(g) + Br_2(l) \text{ – brown fumes}$$
$$8HI(g) + H_2SO_4(l) \rightarrow 4H_2O(l) + H_2S(g) + 4I_2(s) \text{ – purple vapour}$$

2   they would thermally decompose:

$$2HBr(g) \rightleftharpoons H_2(g) + Br_2(g) \text{ and } 2HI(g) \rightleftharpoons H_2(g) + I_2(g)$$

→ Hydrogen bromide is formed when hydrogen burns in bromine vapour: $H_2(g) + Br_2(g) \rightarrow 2HBr(g)$.

It is not energetically feasible to prepare hydrogen iodide by direct combination of iodine with hydrogen – the reaction is endothermic. On the contrary, if you plunge a red hot silica rod into HI(g), you will see a purple vapour and a black precipitate (of iodine) form instantly.

Hydrochloric acid

→ Hydrochloric acid is a solution of $H_3O^+(aq) + Cl^-(aq)$.

This non-oxidizing mineral acid is typical of the 'hydrohalidic' acids in their reactions as aqueous acids with alkalis, basic oxides, carbonates and hydrogencarbonates:

$$HCl(aq) + KOH(aq) \rightarrow KCl(aq) + H_2O(l)$$
$$2HCl(aq) + MgO(s) \rightarrow MgCl_2(aq) + H_2O(l)$$
$$2HCl(aq) + Na_2CO_3(s) \rightarrow 2NaCl(aq) + H_2O(l) + CO_2(g)$$

Halides

In organic compounds, halogen atoms are usually attached to carbon atoms by covalent bonds. Many inorganic halides are covalent or highly covalent in character. Binary s-block metal halides and the halides of d-block metals in their lower oxidation states are usually ionic or highly ionic in character.

- → Identifying a halide means converting it to hydrogen halide, HX(g), or aqueous halide ion, X⁻(aq), and testing it.
- → You should know how to identify chloride, bromide and iodide in theory *and* practice but a fluoride in theory only.

Identifying solid halides

- → *General procedure*: warm a small sample with conc. $H_2SO_4(l)$
- → *General observation*: steamy fumes give white smoke with $NH_3(g)$

Halide	Specific observations
fluoride	fumes etch glass not protected by candle wax
chloride	no colour
bromide	brown colour
iodide	purple colour and grey-black solid

- → *Additional test*: mix halide with manganese(IV) oxide before adding conc. $H_2SO_4(l)$ and identify $Cl_2(g)$ with moist blue litmus
- → *Observation*: litmus paper turns from blue → red → white

Identifying aqueous halide ions

- → *General procedure*: acidify with $HNO_3(aq)$, add aqueous silver nitrate, $AgNO_3(aq)$, then make alkaline with aqueous ammonia
- → *General observation*: precipitate forms and then disappears

Halide	Specific observations
fluoride	no precipitate forms
chloride	white ppt complete 'dissolves' in ammonia
bromide	off-white ppt partly 'dissolves' in ammonia
iodide	pale yellow ppt not affected by ammonia

- → *Additional test*: add aqueous calcium chloride
- → *Observation*: white precipitate (of calcium fluoride)

Interhalogen compounds

Halogens combine with one another to form covalent compounds. All six possible diatomic combinations (ClF, BrF, IF, BrCl, ICl, IBr) exist and are named as the halide of the less reactive halogen.

Action point

Look up and tabulate the lengths and strengths of the carbon–halogen bonds. Compare the data with that for the hydrogen–halogen bonds.

Links

See page 33: Examination question.

Checkpoint 3

Suggest why
(a) we must (i) acidify the solution and (ii) use nitric acid to do so
(b) AgF is soluble in water but AgCl, AgBr and AgI are insoluble and increasingly so from AgCl to AgI
(c) calcium fluoride is insoluble in water even though most ionic fluorides are soluble

Watch out!

You may add drops of $Cl_2(aq)$ to NaBr(aq), add an organic solvent and *not* see a red-brown colour because the displaced $Br_2(aq)$ reacts with the $Cl_2(aq)$ to form BrCl(aq)!

Checkpoint 4

What is the name of (a) ICl, (b) ICl₃?

Action point

Revise VSEPR theory by drawing 'dot and cross' diagrams and predicting the shapes of some interhalogen molecules such as ICl_3, BrF_5 and I_2Cl_6. Hint: the structure but not the shape of I_2Cl_6 is similar to Al_2Cl_6 – the eight atoms in I_2Cl_6 are coplanar.

Links

See page 15: Shapes of molecules.

Exam practice (5 minutes)   answers: page 138

(a) Write an equation for the reaction of (i) chlorine with aluminium and (ii) aqueous iodide ions with manganate(VII) ions in excess acid.

(b) Use the equation $IO_3^-(aq) + 5I^-(aq) + 6H^+(aq) \rightarrow 3I_2(aq) + 3H_2O(l)$ to calculate the concentration of 20 cm³ $KIO_3(aq)$ that liberates 0.15 mol of aqueous iodine molecules from excess potassium iodide solution.

Industrial chemistry: aluminium and carbon

Aluminium, oxygen and silicon are major constituents of many clays and rocks composing the earth's crust. Carbon is the major component of living things. Nitrogen and phosphorus are essential for plant growth and major constituents of fertilizers. Sulphuric acid is such an important industrial chemical that its annual production has been taken as an indication of a nation's economic prosperity.

The jargon

A *mineral* is any substance that can be mined or that occurs naturally as a product of inorganic processes.

Action point

What important factors must be considered when choosing where to build a new industrial aluminium extraction and processing plant?

Grade booster

If you want the top grade, be prepared for questions to test your chemistry in a relevant social, technological or environmental context. For example, could you suggest ways to make industrial chemical processes 'greener'?

Group III

Aluminium

Aluminium is an important 21st century metal. Every year worldwide 22 million tonnes are produced and a further 7 million tonnes are obtained by recycling. An American (Charles Hall) and a Frenchman (Paul Héroult) independently invented in the 19th century the method we still use today to extract the metal from its main source – the mineral *bauxite*, a hydrated oxide, $Al_2O_3 \cdot 2H_2O$.

Extraction of aluminium

Bauxite is mined mainly in Africa, Australia, South America and the West Indies.

Basic Fe_2O_3 and TiO_2 impurities in bauxite are removed from the *amphoteric* Al_2O_3 using sodium hydroxide at 150 °C and 4 atm to form a solution of sodium hydroxyaluminate and insoluble impurities:

$$Al_2O_3/Fe_2O_3/TiO_2(s) + 2NaOH(aq) + 3H_2O(l)$$
$$\rightarrow 2NaAl(OH)_4(aq) + Fe_2O_3/TiO_2(s)$$

After the impurities are removed, the solution is cooled and seeded with pure alumina to precipitate alumina trihydrate which decomposes on heating to 1 000 °C to give pure aluminium oxide.

The jargon

Cryolite is a mineral ore found in Greenland. But Na_3AlF_6 used in the Hall–Héroult process is now synthetic.

Watch out!

Potentially it is a very reactive metal but it has a tough oxide coating that protects the surface from attack.

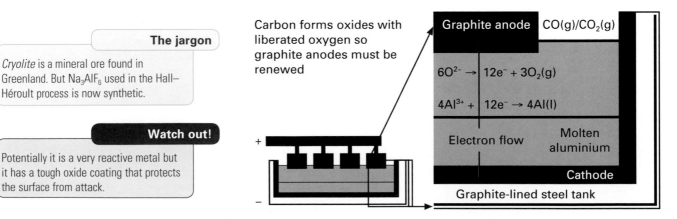

Carbon forms oxides with liberated oxygen so graphite anodes must be renewed

Graphite anode CO(g)/CO₂(g)

$6O^{2-} \rightarrow 12e^- + 3O_2(g)$

$4Al^{3+} + 12e^- \rightarrow 4Al(l)$

Electron flow Molten aluminium

Cathode

Graphite-lined steel tank

Action point

Find out what elements are alloyed with aluminium and what the alloys are used for.

→ Alumina is dissolved in molten cryolite, $(Na^+)_3AlF_6^{3-}$, containing some calcium fluoride to lower the melting point to about 950 °C.
→ Replacement graphite anodes are usually made at the plant.
→ Small amounts of cryolite must be added occasionally.

Aluminium can be anodized and alloyed with other elements to produce low-density materials for specialist uses.

Aluminium compounds

Aluminium oxide: major industrial uses are in glass, paper and pottery manufacture, in water treatment, as a refractory, an abrasive and insulator, and as a catalyst promoter and support in the chemical and petroleum industries.

→ Anodizing is an electrolytic process to thicken the aluminium oxide layer so it adsorbs dyes and permanently colours the metal surface.

Aluminium sulphate: very effective coagulating agent (Al^{3+}(aq) ions) used for water treatment but suspected of causing brain damage.

Aluminium triethyl, $Al(C_2H_5)_3$: used with titanium(IV) chloride in Ziegler–Natta catalysts to produce high-density poly(ethene) and poly(propene).

Lithium aluminium hydride, $LiAlH_4$: a versatile organic reagent used in dry ether to reduce, for example, an unsaturated carboxylic acid to a primary alcohol without affecting the carbon–carbon double bond.

Group IV

Carbon

→ Coke is an essential raw material in the iron and steel industry.
→ Diamonds are used in industrial drills and cutting tools.
→ Graphite is used as a lubricant, as electrodes in industrial electrolytic extraction processes and in pencils.
→ Carbon fibres are used to produce lightweight but very strong materials incorporated into racing cars, tennis rackets, etc.

Carbon compounds

Carbon monoxide: dangerous pollutant (from incomplete combustion of petrol and other hydrocarbons), reducing iron oxides to iron in the blast furnace, combines with hydrogen (when heated under pressure with ZnO/Cr_2O_3 catalysts) in the industrial production of methanol.

Carbon dioxide: acidic gas, produced by combustion of carbon-containing fuels, converted (with water) photosynthetically by plants into carbohydrates, increases now causing the 'greenhouse effect'.

Carbonates: sodium carbonate (Na_2CO_3) and calcium carbonate ($CaCO_3$) are important industrial chemicals.

Hydrocarbons: natural gas (methane), petroleum, etc., used as fuels and feedstock in the petrochemical industry.

Interstitial carbides: non-stoichiometric compounds of carbon with metals like tungsten carbide, extremely hard and used in drill bits.

The jargon

Coagulating agents are chemicals that cause colloidal particles (too small to settle under gravity) to clump into larger particles that can settle or be filtered.

Action point

Find out what happened when a lorry load of aluminium sulphate was dumped into the wrong tank at the Camelford water works.

The jargon

The systematic name for $LiAlH_4$ is *lithium tetrahydridoaluminate(III)*.

Checkpoint

Suggest why $LiAlH_4$ must be used in dry ether.

Action point

Top grade candidates should check out the reductions of carboxylic acids, carbonyl compounds and nitriles with lithium tetrahydridoaluminate(III)

Links

See pages 170: petrochemical industry.

Exam practice (5 minutes) answers: page 138

(a) For the manufacture of aluminium, suggest (i) how the Fe_2O_3 and TiO_2 impurities would be removed from the hot aqueous sodium hydroxyaluminate, (ii) what would happen if the bauxite were not purified before being electrolyzed and (iii) why the cryolite must be replenished.

(b) State how and explain why aluminium is anodized.

Industrial chemistry: silicon and nitrogen

The jargon

Si = silicon (element)
SiO_2 = silica (silicon dioxide)
Zone melting = zone refining

Checkpoint 1

(a) What is the oxidation number of Si in its compounds?
(b) How does the conductivity in a (*n*- or *p*-type) semiconductor change with a rise in temperature?

The formation of ammonia in the Haber process and its subsequent oxidation are very important chemical processes. Ammonia is needed for fertilizer production, which uses 80% of all NH_3 made, and the production of nitric acid, explosives and nylon.

Silicon

The element is extracted by heating quartz in a carbon arc:

$$SiO_2 + 2C \rightarrow Si + 2CO$$

Very pure silicon is produced by

1. making silicon chloride: $Si(s) + 2Cl_2(g) \rightarrow SiCl_4(l)$
2. purifying the chloride by fractional distillation
3. reducing the chloride using very pure magnesium:
   $SiCl_4(l) + 2Mg(s) \rightarrow Si(s) + 2MgCl_2(s)$
4. purifying the element by zone melting

Action point

Find out what silicones are and why they are widely used as lubricants, elastomers, water-repellent oils and waxes. Hint: elastomers are synthetic rubbers.

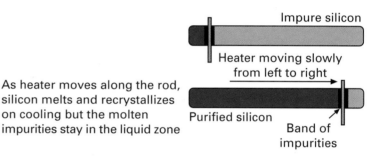

As heater moves along the rod, silicon melts and recrystallizes on cooling but the molten impurities stay in the liquid zone

Impure silicon

Heater moving slowly from left to right

Purified silicon

Band of impurities

The production of extremely pure semiconducting silicon crystals led to silicon chips, modern electronics and computer technology.

The jargon

Clay is a complicated mixture of silicates and aluminosilicates.
Portland cement is so-called because its inventor thought it looked like the stone from the quarries in Portland, Dorset.

Silicates

→ Glass is a transparent mixture of silicates manufactured by melting sandstone, limestone and soda in a high-temperature furnace.

The complex reactions may be represented by two simplified equations:

$$CaCO_3(s) + SiO_2(s) \rightarrow CaSiO_3(l) + CO_2(g)$$
$$Na_2CO_3(s) + SiO_2(s) \rightarrow Na_2SiO_3(l) + CO_2(g)$$

Checkpoint 2

Why is the use of blue asbestos prohibited and the use of white asbestos carefully controlled?

→ Cement is a complex mixture manufactured by heating limestone and clay in a rotary kiln, grinding and then mixing with gypsum.

The complex reactions may be represented by a simplified equation:

$$CaO \cdot Al_2O_3 \cdot 2SiO_2(s) + 8CaCO_3(s) \rightarrow 2Ca_3SiO_5(s) + Ca_3Al_2O_6(s) + 8CO_2(g)$$

The setting of cement is a complex exothermic process involving water stirred into the powder and carbon dioxide absorbed from the air.

→ Asbestos is an industrially useful heat-resistant fibrous mineral.

Group V

→ Nitrogen is produced by fractional distillation of liquid air.
→ Nitrogen is the raw material for the large-scale industrial production of ammonia and nitric acid.

Manufacture of ammonia by the Haber process

→ Nitrogen (with argon) comes from the air.
→ Hydrogen may come from water gas, steam reforming of natural gas, or catalytic reforming processes in the petroleum industry:

$$N_2(g) + 3H_2(g) \xrightleftharpoons[\text{promoted iron}]{\text{450°C 200 atm}} 2NH_3(g); \Delta H = -92 \text{ kJ mol}^{-1}$$

catalyst

→ The heterogeneous catalyst is made by putting a mixture of oxides of iron, potassium, calcium and aluminium into the reaction vessel.

According to the **law of chemical equilibrium** formation of ammonia is favoured by high pressure and low temperature. According to the **kinetic molecular theory** rate of formation is favoured by high temperature and catalysts. In practice,

→ 15% conversion is achieved under optimum conditions
→ gases from the reaction vessel are cooled while still under pressure
→ liquid ammonia is removed as it forms

The major industrial use of ammonia is in the manufacture of fertilizers, nylon and nitric acid.

Manufacture of nitric acid by the Ostwald process

$$4NH_3(g) + 5O_2(g) \rightleftharpoons 4NO(g) + 6H_2O(g); \Delta H = -994 \text{ kJ}$$

5 atm pressure Pt/Rh gauze ← keeps gauze catalyst hot
at 900°C

The gaseous product is compressed, cooled and mixed with air so the nitrogen oxide can spontaneously oxidize to nitrogen dioxide and hydrate to nitric acid containing around 60% HNO_3 by mass:

$$2NO(g) + O_2(g) \rightarrow 2NO_2(g)$$
$$4NO_2(g) + O_2(g) + 2H_2O(l) \rightarrow 4HNO_3(aq)$$

The major industrial use of nitric acid is the production of ammonium nitrate (as an explosive in quarrying and a fertilizer in agriculture) and in the **nitration** of organic compounds to produce explosives such as nitroglycerine, nitrocellulose and trinitrotoluene.

The jargon

Natural gas = methane
Water gas = $CO(g) + H_2(g)$ formed by passing steam over white-hot coke
Catalytic reforming: see page 162
Steam reforming = $CH_4(g) + H_2O(g)$ under pressure over heated Ni catalyst to form $CO(g) + 3H_2(g)$
Promoted iron catalyst = finely divided iron mixed with oxides of K, Ca and Al to enhance its activity

Action point

See page 176 for recent developments including the KAAP process that uses a graphite supported ruthenium catalyst with 20 times the activity of traditional iron catalyst.

Checkpoint 3

What happens to the 85% unconverted hydrogen and nitrogen?

Action point

Nitrocellulose and potassium nitrate are the main ingredients of *Pyrogen* - a system that explosively produces a fire extinguishing aerosol.
Visit http://www.pyrogen.co.uk

Jargon

An *aerosol* is a suspension of tiny (usually sub-microscopic) solid particles and/or liquid droplets in a gas (usually air).

Action point

Heterogeneous catalysis is very important in industrial chemistry. Look through your specification and note any industrial processes that involve heterogeneous catalysts. You will find examples in your organic chemistry as well as in this inorganic chemistry section.

Exam practice (10 minutes)   answers: page 138

The Haber process *fixes nitrogen* by synthesizing ammonia from nitrogen and hydrogen *under optimum conditions* using iron as a *heterogeneous catalyst*.

(a) Explain what is meant by the terms in italics.
(b) For the synthesis of ammonia, write an expression for K_p and use it to explain (i) why the Haber process operates at high pressure and (ii) why the gases are cooled while still under pressure and liquid ammonia is removed continuously as it forms from the reaction vessel.
(c) (i) Suggest why the catalyst is made in the reaction vessel by reducing iron(III) oxide and (ii) write a balanced equation for the reduction.

Industrial chemistry: sulphur and the halogens

Sulphur is extracted from the ground by the Frasch process or by desulphurization of fossil fuels, and used to produce sulphuric acid. Sodium chloride is mined or extracted from salt deposits as brine and electrolyzed to produce chlorine and sodium hydroxide.

Checkpoint 1

(a) Suggest what would happen if the gases were not (i) purified, (ii) cooled in heat exchangers.
(b) What is meant by (i) heterogeneous catalyst, (ii) promoter?

The Contact process for sulphuric acid

→ Sulphur dioxide comes from burning sulphur in air or from roasting sulphide minerals in some metal extraction plants:

$$S(s) + O_2(s) \rightarrow SO_2(g); \quad ZnS(s) + 1\frac{1}{2}O_2(s) \rightarrow ZnO(s) + SO_2(g)$$

→ Sulphur dioxide and excess air are purified and passed through a series of alternate reaction vessels and heat exchangers:

$$2SO_2(g) + O_2(g) \underset{\text{at 430 °C}}{\overset{V_2O_5 \text{ catalyst}}{\rightleftharpoons}} 2SO_3(g); \quad \Delta H = -197 \text{ kJ}$$

removed by heat exchangers

→ The reaction vessels hold a heterogeneous catalyst of vanadium(V) oxide promoted by potassium sulphate and supported on silica.
→ A heat exchanger cools the gases to about 430 °C before they enter a reaction vessel to form sulphur trioxide exothermically.

The jargon

Mist is liquid droplets suspended in air. *Oleum* (fuming sulphuric acid) is a 'solution' of $SO_3(s)$ in $H_2SO_4(l)$.

In modern plants, conversion of SO_2 to SO_3 is almost complete.

→ Sulphur trioxide gas is absorbed in 98% sulphuric acid to form oleum ($H_2S_2O_7$) as the final product or for dilution to $H_2SO_4(l)$.
→ Sulphur trioxide gas is *not* added to water alone because that would produce an inconvenient stable acid mist.

Grade booster

This is a favourite topic in exams. You must be able to explain the chemical principles behind the Contact process if you want the top grade.

Any unreacted sulphur dioxide is passed through an extra converter as an environmental control to minimize SO_2 emissions.

The optimum operating conditions for the process follow from the law of chemical equilibrium and the principles of reaction kinetics.

$$\frac{p^2_{SO_3}}{p^2_{SO_2} \times p_{O_2}} = K_p$$

High pressure would improve the equilibrium yield of sulphur trioxide and increase its rate of formation. In practice the slight improvement would not justify the expense of high-pressure plant.
Low temperature would improve the equilibrium yield of sulphur trioxide but decrease its rate of formation.
Heterogeneous catalyst would not improve the equilibrium yield of sulphur trioxide but would increase its rate of formation.
Optimum conditions

Links

See page 53: law of chemical equilibrium.

Checkpoint 2

Explain how the law of chemical equilibrium predicts that the yield of SO_3 in this reversible reaction would improve at high pressure and low temperature.

→ enough pressure (about 2 atm) to pump gases through the reaction vessels and heat exchangers
→ compromize temperature to achieve a reasonable yield at a reasonable rate
→ solid catalyst to speed up the rate of attainment of equilibrium

Group VII

Manufacture of chlorine

Chlorine is produced by the electrolysis of chlorides during the manufacture of sodium hydroxide and reactive metals. Two important electrolytic processes may be summarized as follows:

$$NaCl(aq) + H_2O(l) \xrightarrow{\text{D.C. electricity}} \underbrace{NaOH(aq)}_{\text{at the cathode}} + H_2(g) + \underbrace{Cl_2(g)}_{\text{at the anode}}$$

Links

See page 96: sodium hydroxide and chlorine manufacture.

Uses of chlorine

→ water purification
→ manufacture of PVC, organic chemicals and chlorinated solvents
→ manufacture of hydrochloric acid and inorganic chemicals such as $NaClO(aq)$ and $NaClO_3(s)$

Manufacture of bromine and iodine

→ Bromine is mostly extracted from acidified sea water by oxidizing bromide ions with chlorine in a non-metal displacement reaction:

$$Cl_2(g) + 2Br^-(aq) \rightarrow 2Cl^-(aq) + Br_2(g)$$

→ Iodine is mostly extracted by reducing sodium iodate(V) from Chile saltpetre (sodium nitrate) deposits using sulphur dioxide:

$$IO_3^-(aq) + 3SO_2(g) + 3H_2O(l) \rightarrow I^-(aq) + 3H_2SO_4(aq)$$
$$IO_3^-(aq) + 5I^-(aq) + 6H^+(aq) \rightarrow 3I_2(s) + 3H_2O(l)$$

Checkpoint 3

What are CFCs and why are they banned in the EU?

Watch out!

For the top grade examiners would expect you to show your knowledge of homolysis of covalent bonds and free radical reactions when they ask you about the effect of CFCs in the stratosphere

Hydrogen chloride

Hydrogen chloride is manufactured by burning hydrogen in chlorine, both gases coming from the electrolysis of $NaCl(aq)$.

Action point

Make a list of the industrial uses of hydrogen chloride and hydrochloric acid.

Exam practice (20 minutes)   answers: page 139

(a) Outline how sulphur dioxide is converted to sulphuric acid on an industrial scale.

(b) Describe and explain what is observed on warming in a test tube a few drops of concentrated sulphuric acid added to (i) $NaCl(s)$, (ii) $KNO_3(s)$, (iii) $KI(s)$, (iv) a mixture of ethanoic acid and pentan-1-ol.

(c) State the main industrial source of (i) bromine and (ii) iodine.

(d) Explain in outline how each halogen in (c) is obtained on a large scale.

(e) Explain briefly how and why aqueous silver nitrate can be used to distinguish aqueous chloride, bromide and iodide but not fluoride ions.

(f) State *two* chemicals manufactured from chlorine on the industrial scale and state one important commercial use for each chemical.

Group VIII (or 0): the noble gases

Action point

Study a detailed account of the discovery of the noble gases and the part played by spectroscopy. The experiments of people like Cavendish, Rayleigh, Ramsay and Travers are a good example of how scientists work.

The jargon

Greek:
Helios – the sun
Neos – new
Argos – idle one
Kryptos – hidden
Xenos – guest, stranger or foreigner

Checkpoint 1

What are the noble metals?

In 1785 Henry Cavendish, an eccentric genius, suspected that less than 1 in 120 parts of our atmosphere is unreactive and different from the rest of the air. More than 100 years later, Rayleigh and Ramsay discovered that air contained about 0.94% of a gas that was more inert than nitrogen itself.

The elements

When these elements were first added to the periodic table they were called the *inert gases* and labelled *group 0*.

	Electronic configuration	E_{m1}/kJ mol^{-1}	M.p./°C	B.p./°C
He	$1s^2$	2 372	−272	−269
Ne	$1s^2 2s^2 2p^6$	2 080	−249	−246
Ar	$1s^2 2s^2 2p^6 3s^2 3p^6$	1 519	−189	−186
Kr	$1s^2 2s^2 2p^6 3s^2 3p^6 3d^{10} 4s^2 4p^6$	13 511	−157	−152
Xe	$[Kr]4d^{10}5s^2 5p^6$	1 170	−112	−108
Rn	$[Xe]4f^{14}5d^{10}6s^2 6p^6$	1 037	−71	−62

Helium

Janssen found evidence of an undiscovered element (helium) during the solar eclipse in 1868 when he observed a new line in the sun's spectrum. In 1870 Sir Norman Lockyer suggested the line was due to an element that he named 'Helium', after the Greek Sun god 'Helios'. In 1895 Ramsay discovered helium in *clevite*, a uranium mineral. Two Swedish chemists, Cleve and Langlet, also made the same discovery independently of Ramsay.

→ Helium is obtained by the fractional distillation of liquid air.
→ Most commercial helium comes from natural gas and petroleum.

Helium in the earth's crust is from α decay of heavy nuclides.

→ An α particle is a nucleus of two protons and two neutrons.
→ A helium atom forms when an α particle gains two electrons.

Argon (and neon, krypton and xenon)

→ Argon is the most abundant noble gas ($\approx$ 1%) in the atmosphere.
→ Argon (neon, krypton and xenon) are separated from the atmosphere by fractional distillation of liquid air.

Radon

→ Radon is a radioactive alpha emitter formed in the radioactive decay of heavy nuclides such as ^{238}U.

Over thousands of years radon has accumulated in certain rocks. In some places the gas has been detected inside houses and its presence has been linked to the incidence of some types of cancers.

→ The radon is thought to seep into houses from the rocks and to be a health risk.
→ One remedy is increased interior and under-floor ventilation to minimize the concentration of radon gas inside the house.

Uses of the noble gases

Helium

→ weather balloons and airships
→ underwater breathing apparatus (80% helium and 20% oxygen)
→ low temperature physics and low temperature research

Argon

→ largest use as an *inert atmosphere* in arc cutting and welding, in titanium production, in silicon and germanium zone refining, and in electric light bulbs to inhibit evaporation of the filaments

Neon, krypton and xenon

→ different low pressure mixtures of these three gases are used to fill fluorescent tubes and produce different coloured lights

Atomic emission spectroscopy

An electrical discharge through gases or vapours at low pressures excites the atoms from their ground state to higher energy levels. The energy emitted upon return to the ground state results in characteristic narrow line spectra (often as visible colours) for each element. The intensity of the emitted light is proportional to the concentration of the atoms. An AES (high-resolution atomic emission spectrometer and polychromator (wavelength selector) with multiple detectors) is used in chemical analysis to detect and determine elements in sample.

sample injection → excitation source → high-resolution polychromator → multiple detector → recorder

The chemistry of the noble gases

Until 1962 chemists had not made any true noble gas compound although they thought it should be possible to make $XePtF_6$ in view of the similarity of the first ionization energy of the oxygen *molecule* ($E_{m1} = 1175$ kJ mol^{-1}) and xenon *atom* ($E_{m1} = 1170$ kJ mol^{-1}) and the existence of $O_2^+PtF_6^-$.

→ In 1962, Neil Bartlett reported the preparation of $XePtF_6$.

Within a few months of Bartlett's announcement, chemists had prepared xenon fluorides by direct combination of the two elements.

→ Xenon and fluorine combine at 6 atm pressure when passed through a nickel tube heated to 400 °C:

$$Xe(g) + F_2(g) \rightarrow XeF_2(s)$$
$$Xe(g) + 2F_2(g) \rightarrow XeF_4(s)$$
$$Xe(g) + 3F_2(g) \rightarrow XeF_6(s)$$

The three xenon fluorides are all colourless crystals. We can store them indefinitely in nickel vessels but the higher fluorides are hydrolysed by a trace of water. The fluorides hydrolyse to xenon oxofluorides and xenon oxides. The majority of noble gas compounds are the compounds of xenon. Krypton difluoride and radon difluoride have been prepared. Ever since 1962 when Bartlett made a compound of Xenon, we have called Group VIII in the periodic table the noble gases.

Exam practice (8 minutes)   answers: page 139

(a) Write the ground state electronic configuration, in terms of s and p electrons, for an isolated argon atom and use it to explain the stability of chloride and potassium ions relative to chlorine gas and potassium metal.

(b) State how XeF_2 and XeF_4 may be prepared and deduce the shapes of the molecules.

Watch out!

For the top grade you should remember how to do a simple flame test and know that the common visible colours are

sodium	golden yellow
potassium	lilac
lithium	red
strontium	crimson
barium	apple green
copper	bluish green

Sodium vapour lamps are used for airfield and street lighting where colour distinction is not so important. The are red during the 30 minute warm up period. Mercury vapour lamps gives a bluish colour and are used in street lighting. They also emit UV light making them suitable in vivaria for reptile species known to bask in full sunshine

Action point

Find out the difference between atomic emission and atomic absorption spectra.

Checkpoint 2

(a) What is meant by the molar first ionization energy of the oxygen molecule?

(b) Explain briefly in terms of energetics why it may be impossible to combine fluorine with helium or neon.

Watch out!

$E_{m1} = 1314$ kJ mol^{-1} for atomic oxygen.

Links

See page 126: d-block catalysts.

1st transition series: metals, aqueous ions and redox

The colour and variety of oxidation states of the transition elements make them a fascinating area of study but quite tricky at times!

The jargon

Sc–Zn = first series of ten d-block elements with inner d-orbitals filling
Ti–Cu = eight metals with incomplete d-orbitals in some compounds
Sc, Zn = two metals with no incomplete d-orbitals in any compound.

The elements

Physical properties and uses of the metals

Unlike the physical properties of elements in a row across the p-block, the typical metallic properties of these d-block elements show little or no change with increasing atomic number from scandium to zinc.

→ They are all ductile and malleable, lustrous and sonorous hard metals with high m.p.s, b.p.s, tensile strength and electrical conductivity.
→ They readily form *alloys* and *interstitial compounds* because they have similar properties and atom size.

You should tie the above properties of d-block elements to some of their main uses, e.g. electrical and thermal conductivity for electrical wire (copper) and heating elements (nichrome), lustre and hardness for coins and chrome plating, sonority for musical instruments, etc.

→ The key to the characteristic physical and chemical properties of transition metals (Ti–Cu) is their ground state electronic configurations and *incomplete 3d-subshells*

Watch out!

Only one 4s electron for Cr and Cu.

Grade booster

For the top grade, you should be able to explain that the ground state configuration for Cr and Cu is $4s^1$ (not $4s^2$) because five half-full d-orbitals and five full d-orbitals are particularly stable arrangements.

Links

See page 11: electronic configurations.

Links

See page 46: catalysis.

Checkpoint 1

Write an equation for (a) sodium reducing titanium(IV) chloride to titanium, (b) the extraction and refining of nickel using carbon monoxide, and (c) the ion–electron half-reaction for the reduction of copper(II) ions at the cathode during electrolytic refining.

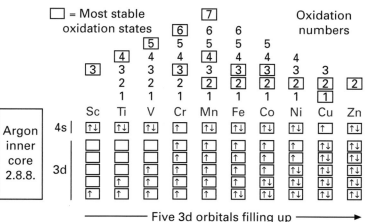

→ Transition elements are efficient *heterogeneous catalysts*.

For example, iron (in the Haber synthesis of ammonia) and nickel (in the hydrogenation of unsaturated oils) can use their d-orbitals when adsorbing the reacting gases to lower the activation energy barrier.

Extraction of the transition metals

→ The metals are usually obtained by chemical or electrolytic reduction of their mineral ores, e.g. iron in the blast furnace:

$$Fe_2O_3(s) + 3CO(g) \rightleftharpoons 2Fe(l) + 3CO_2(g)$$

The metals are purified by various methods.

→ Copper is refined electrolytically and titanium by a process involving decomposition of TiI_4 on a hot wire.

Chemical properties

→ The metals form a variety of compounds with non-metals and react to varying extents with water and acids: e.g.

$$3Fe(s) + 4H_2O(g) \rightarrow Fe_3O_4(s) + 4H_2(g)$$
$$Fe(s) + 2HCl(g) \rightarrow FeCl_2(s) + H_2(g)$$
$$2Fe(s) + 3Cl_2(g) \rightarrow 2FeCl_3(s)$$

→ Rusting (and its prevention by a sacrificial metal) involves electrochemical reactions that include oxidation of iron (and of the sacrificial metal) and reduction of oxygen: e.g.

$$Fe(s) \rightarrow Fe^{2+}(aq) + 2e^-, \text{ then } Fe^{2+}(aq) \rightarrow Fe^{3+}(aq) + e^-$$
$$Zn(s) \rightarrow Zn^{2+}(aq) + 2e^-, \text{ then } O_2(g) + 2H_2O(l) + 4e^- \rightarrow 4OH^-(aq)$$

Watch out!

In these reactions the gases are passed over heated iron which must be kept red-hot except in the third case when the metal 'burns' in chlorine.

The transitional metal ions

The characteristic features of transition metal compounds are

→ variety and relative stability of oxidation states
→ formation and stability of various complexes
→ ability to act as catalysts

Grade booster

Favourite questions are about redox reactions, complex ion formation and homogeneous catalysis involving aqueous transition metal ions. The most popular elements are Mn, Cr, Fe and Cu. Be prepared if you want the top grade.

These features depend on d-orbitals being used for bonding.

For A-level the emphasis is on the *aqueous metal ions*. Most aqueous solutions of d-block compounds have distinctive colours associated with transition metal cations and/or oxoanions. You probably already associate blue with copper(II) sulphate, $Cu^{2+}(aq)$ and purple with potassium manganate(VII), $MnO_4^-(aq)$.

→ Colours of aqueous solutions change when transition metal ions take part in electron-transfer or ligand-transfer reactions.

Links

See page 164: UV and visible spectroscopy.

We explain the colours of these aqueous solutions by electrons in the transition metal ions changing d-orbitals. The associated energy changes result in light (energy) of different colours being absorbed.

Redox reactions

Now is a good time to revise oxidation number rules, ion–electron half-equations, electrochemical cells and standard electrode potentials. We interpret a redox reaction of aqueous transition metal ions as an electron transfer and use $E^{\ominus}$ values to predict its feasibility.

For the elements Sc to Zn:

→ Stability of the higher oxidation states decreases from Sc to Zn.
→ Stability of the +2 oxidation state relative to the +3 oxidation state increases from Sc to Zn.
→ Greater stability of Mn^{2+} with respect to Mn^{3+} and Fe^{3+} with respect to Fe^{2+} may be related to the stability of the half-filled d-subshell.
→ Ions tend to exist as aqua cations in low oxidation states and oxoanions in higher oxidation states.

Checkpoint 2

Write the formulae of (a) the chlorides and (b) the sulphates of the first row of d-block elements in their most stable oxidation states.

Exam practice (8 minutes)   answers: page 140

(a) Explain briefly why zinc may be regarded as a d-block element but not a transition metal.

(b) State and account for the colour changes in solution when an excess of granulated zinc is added to aqueous sulphuric acid containing (i) $K_2Cr_2O_7(aq)$ and (ii) $NH_4VO_3(aq)$.

1st transition series: redox reactions and

Transition metal ions undergo a variety of redox reactions, form a range of complexes and catalyze many reactions.

The jargon

Titrimetric analysis =

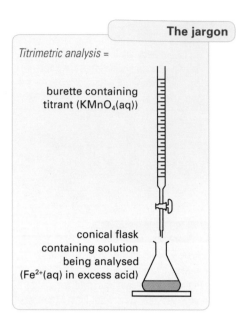

burette containing titrant (KMnO₄(aq))

conical flask containing solution being analysed (Fe²⁺(aq) in excess acid)

Redox titrations

Potassium manganate(VII) and potassium dichromate(VI) are important oxidizing agents in titrimetric analysis. When they are used with excess sulphuric acid the reactions are quantitative and their reduction is represented by the following ion–electron half-equations:

$$MnO_4^-(aq) + 8H^+(aq) + 5e^- \rightarrow Mn^{2+}(aq) + 4H_2O(l)$$
$$Cr_2O_7^{2-}(aq) + 14H^+(aq) + 6e^- \rightarrow Cr^{3+}(aq) + 7H_2O(l)$$

When you run aqueous potassium manganate(VII) solution into aqueous iron(II) sulphate in excess sulphuric acid, the purple colour disappears as the iron(II) ions are oxidized to iron(III) ions:

$$Fe^{2+}(aq) \rightarrow Fe^{3+}(aq) + e^-$$

The oxidation number of manganese falls from +7 to +2 (5 units down) and that of iron rises from +2 to +3 (1 unit up).

1 mol MnO_4^-(aq) must oxidize 5 mol Fe^{2+}(aq) to balance the e⁻s

We can combine the reduction and oxidation half-equations to eliminate the electrons and get the stoichiometric redox equation:

$$MnO_4^-(aq) + 8H^+(aq) + 5Fe^{2+}(aq) \rightarrow Mn^{2+}(aq) + 4H_2O(l) + 5Fe^{3+}(aq)$$

We could use this reaction in an electrochemical cell:

$$Pt\,|\,[Fe^{2+}(aq),Fe^{3+}(aq)]\,\vdots\vdots\,[MnO_4^-(aq) + 8H^+(aq)],[Mn^{2+}(aq) + 4H_2O(l)]\,|\,Pt$$

and use the standard electrode potentials to confirm that the reaction is energetically feasible:

Redox process	$E^\ominus$/V
$MnO_4^-(aq) + 8H^+(aq) + 5e^- \rightleftharpoons Mn^{2+}(aq) + 4H_2O(l)$	+1.51
$Fe^{3+}(aq) + e^- \rightleftharpoons Fe^{2+}(aq)$	+0.77
$MnO_4^-(aq) + 8H^+(aq) + 5Fe^{2+}(aq)$	
$\rightarrow Mn^{2+}(aq) + 4H_2O(l) + 5Fe^{3+}(aq)$	+0.74

$E^\ominus$ is positive (+0.74 V) so the reaction is energetically feasible.

Homogeneous catalysis

Many aqueous transition metal ions act as homogeneous catalysts. We might expect the reaction between peroxodisulphate(VI) ions and iodide ions to be slow because both are anions. The reaction is catalysed by iron(II) cations or iron(III) cations. You could explain how the catalyst works by taking part in the redox reaction to provide an alternative route with a lower activation energy.

Iron(III) ions rapidly oxidize the iodide ions to iodine:

$$2Fe^{3+}(aq) + 2I^-(aq) \rightarrow 2Fe^{2+}(aq) + I_2(aq) \qquad E^\ominus = 0.23\ V$$
$$2Fe^{2+}(aq) + S_2O_8^{2-}(aq) \rightarrow 2Fe^{3+}(aq) + 2SO_4^{2-}(aq) \qquad E^\ominus = 1.24\ V$$

Iron(II) ions are rapidly oxidized by the peroxodisulphate(IV) ions.

Combine the two steps involving the catalyst to get the equation

Grade booster

If you want the top grade, be prepared for redox titrations you met in your coursework and practical work to appear in questions in your theory papers.

Checkpoint

Calculate the standard free energy change for the redox reaction using the value +0.74 V and the expression
$$\Delta G^\ominus = -zFE^\ominus \text{ (see page 83)}$$

Links

See pages 46–7: homogeneous catalysis.

complex ion formation

$$S_2O_8^{2-}(aq) + 2I^-(aq) \rightarrow 2SO_4^{2-}(aq) + I_2(aq)$$

The catalyst takes part in the reaction but is not used up.

Complex formation

→ A complex ion is a central metal cation with six (four or two) ligands (molecules or ions) datively bonded to it.

→ The nucleophilic molecules CO, H_2O, NH_3 and nucleophilic ions OH^-, CN^-, F^-, Cl^- are monodentate ligands.

→ 1,2-diaminoethane $NH_2CH_2CH_2NH_2$ and the ethanedioate ion $^-O_2C \cdot CO_2^-$ are bidentate ligands.

→ EDTA, ethylenediaminetetraacetic acid, is a hexadentate ligand consisting of many atoms, six of which are each capable of forming one dative bond with the central ion.

The shapes of complex ions may sometimes (not always) be predicted using VSEPR theory. Octahedral is most common and can give rise to geometrical and optical isomeric complexes. There are also linear, square planar and tetrahedral complexes.

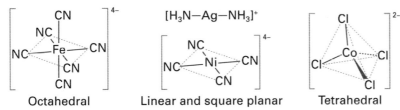

| Octahedral | Linear and square planar | Tetrahedral |

→ Reactions can occur with one ligand displacing another

When you dissolve copper(II) carbonate in the minimum amount of concentrated hydrochloric acid you get a dirty yellow-brown solution containing $[CuCl_4]^{2-}$ anions. When you add water, the solution changes colour from yellow-brown through green to blue, the colour of an aqueous solution containing $[Cu(H_2O)_6]^{2+}$ cations.

→ Water molecules can replace chloride ions as ligands:

$$[CuCl_4]^{2-}(conc \cdot HCl) + 6H_2O(l) \rightleftharpoons [Cu(H_2O)_6]^{2+}(aq) + 4Cl^-(aq)$$

If you add ammonia to aqueous copper(II) sulphate, the alkali first precipitates copper(II) hydroxide which redissolves in excess ammonia to form a deep blue solution.

→ Ammonia molecules can replace water molecules as ligands:

$$[Cu(H_2O)_6]^{2+}(aq) + 4NH_3(aq) \rightleftharpoons [Cu(NH_3)_4(H_2O)_2]^{2+}(aq) + 4H_2O(l)$$

The jargon

An *anion* is a negatively charged ion. A *cation* is a positively charged ion.

The jargon

A *ligand* is a nucleophile (molecule or anion) datively bonded by a lone electron-pair to a central transition metal cation.
Monodentate means one tooth or dative bond:

$$EDTA = \begin{array}{l} CH_2N(CH_2CO_2H)_2 \\ | \\ CH_2N(CH_2CO_2H)_2 \end{array}$$

EDTA is used to soften water (by sequestering calcium ions) and to reduce blood clotting and the risk of heart attacks.

Links

See pages 15 and 17: VSEPR theory.

Watch out!

The hexaamminecopper(II) ion, $[Cu(NH_3)_6]^{2+}$ is only formed in liquid ammonia and not in aqueous solutions.

Grade booster

For the top grade, make sure your drawing shows clearly the three-dimensional shape of a complex.

Exam practice (12 minutes)　　　　　answers: page 140

(a) Explain, with an example in each case, the meaning of the term
(i) complex cation, (ii) complex anion, (iii) bidentate ligand.

(b) The antidote for cyanide poisoning is to freshly mix $Na_2CO_3(aq)$ with $FeSO_4(aq)$ containing citric acid (2-hydroxypropane-1,2,3-trioic acid) and swallow immediately.
(i) Explain why the two solutions cannot be mixed and stored until needed and suggest how the antidote works.
(ii) Draw the structural formula of citric acid and suggest *two* reasons why it is in the antidote.

1st transition series: chromium

Chromium is a typical d-block element and transition metal. Many chromium compounds are brightly coloured. Some are used in paints because the colours do not fade in the sun.

The jargon

Greek *chroma* – colour

Action point

Find out the colours of the following insoluble chromates:
silver chromate, Ag_2CrO_4
lead(II) chromate, $PbCrO_4$
barium chromate, $BaCrO_4$

The element

Chromium shows all the general characteristics of a transition metal:

→ high melting point (second highest in the first transition series)
→ hard, lustrous and resistant to chemical attack (chromium plating)
→ forms tough alloy with iron (stainless steel)

The Goldschmidt process used aluminium to reduce chromium oxide to the metal in a type of 'thermite' reaction:

$$Cr_2O_3 + 2Al \rightarrow 2Cr + Al_2O_3$$

Oxidation states

Chromium exists in three oxidation states:

→ +2   strongly reducing (spontaneously oxidized by air)
→ +3   most stable as chromium(III) oxide, Cr_2O_3
→ +6   strongly oxidizing (as dichromate(VI) ion, $Cr_2O_7^{2-}$)

Watch out!

Cr has four oxidation numbers if you include 0 for the oxidation state of the uncombined metal.

Chromium(VI) compounds

Chromium(VI) oxide

Chromium(VI) oxide forms when concentrated sulphuric acid is added to cold, concentrated aqueous sodium dichromate(VI). Red crystals of $CrO_3(s)$ separate from the orange sodium dichromate solution.

$$Cr_2O_7^{2-}(aq) + H_2SO_4(l) \rightarrow 2CrO_3(s) + SO_4^{2-}(aq) + H_2O(l)$$

→ Chromium(VI) oxide is a powerful oxidizing agent

Before modern surfactants, 'chromic acid' (a solution of chromium(VI) oxide in concentrated sulphuric acid) was used for cleaning laboratory glassware. It is extremely corrosive, it reacts violently with many organic compounds and it causes skin ulcers. These days chemists prefer to clean glassware with safer modern surfactants.

The jargon

Chromium trioxide is the old name for CrO_3.

Chromates and dichromates

Na_2CrO_4 and $K_2Cr_2O_7$ strictly refer to sodium chromate(VI) and potassium dichromate(VI) but we often leave off the (VI).

→ The oxidation state of chromium is +6 in chromates *and* dichromates.
→ Sodium dichromate is more soluble than potassium dichromate.
→ Aqueous chromate solutions are *yellow* and stable when *alkaline or neutral*.
→ Aqueous dichromate solutions are *orange* and stable when *acidic*.
→ Chromate and dichromate ions are related by the acid–base reaction
$2CrO_4^{2-}(aq) + 2H^+(aq) \rightleftharpoons Cr_2O_7^{2-}(aq) + H_2O(l)$.

This equilibrium explains why insoluble chromates precipitate when some aqueous metal ions are added to aqueous dichromate ions.

Watch out!

If you add $AgNO_3(aq)$ to $K_2Cr_2O_7(aq)$ you will get a precipitate of silver chromate (*not* dichromate).

Grade booster

The equilibrium between aqueous chromate and dichromate ions is acid-base. If you want the top grade, don't lose marks by incorrectly identifying the reversible reaction as a redox. The chromium oxidation state stays unchanged at +6. You may write
$Cr_2O_7^{2-} + 2OH^- \rightleftharpoons 2CrO_4^{2-} + H_2O$
as the acid/base equilibrium.

Chromium(III) compounds

Chromium(III) oxide and hydroxide

→ amphoteric oxide prepared *in a fume cupboard* by starting (heat) the exothermic 'decomposition' of ammonium dichromate:

$$(NH_4)_2Cr_2O_7 \rightarrow Cr_2O_3 + 4H_2O + N_2$$

→ amphoteric hydroxide precipitated by aqueous NaOH(aq) from aqueous chromium(III) salts as a green jelly that dissolves in excess NaOH(aq) to form a green solution.

Chromium(II) compounds

Obtained by reducing chromium(III) compounds (e.g. $Cr^{3+}(aq) \rightarrow Cr^{2+}(aq)$ by acid and zinc amalgam) and keeping the chromium(II) product in an inert atmosphere or under reducing conditions.

Isomerism in chromium compounds

Hydration isomerism

Hydrated chromium(III) chloride, $CrCl_3 \cdot 6H_2O$, can exist as isomers depending upon whether the chlorine is ionic or is a ligand.

$[Cr(H_2O)_4Cl_2]^+Cl^- \cdot 2H_2O$ dark green
$[Cr(H_2O)_6]^3(Cl^-)_3$ grey-blue
$[Cr(H_2O)_5Cl]^{2+}(Cl^-)_2 \cdot H_2O$ light green

Geometrical isomerism

→ octahedral dichlorotetra-ammine chromium(III) ions can exist as geometrical (*cis* and *trans*) isomers

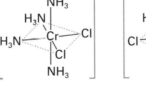

Optical isomerism

→ 1,2-diaminoethane ligands can complex chromium(III) to give optical (*mirror-image*) isomers

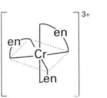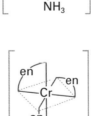

Examiner's secrets

Decompositions are endothermic. This reaction is not a decomposition because the dichromate ion oxidizes the ammonium ion to nitrogen exothermically. It is also very dangerous because dichromate dust becomes airborne!

Watch out!

Take precautions when handling chromium compounds in the lab. Contact with the skin can cause dermatitis.

The jargon

Latin: *cis* – on this side of
trans – on the other side of

Checkpoint

Draw diagrams to show geometric isomers of the square planar complex, $Pt(NH_3)_2Cl_2$.

The jargon

en = $H_2NCH_2CH_2NH_2$ (bidentate ligand)

Exam practice (8 minutes)   answers: page 140

(a) Deduce the oxidation state of chromium in (i) $Cr_2(SO_4)_3$ and (ii) CrO_2Cl_2.

(b) Construct equations for (i) the reaction of aqueous sodium hydroxide with $Cr_2(SO_4)_3$(aq), (ii) the formation of chromium dichloride dioxide from NaCl(s), $K_2Cr_2O_7$(s) and H_2SO_4(l), (iii) the hydrolysis of chromium dichloride dioxide by NaOH(aq) to form aqueous chromate(VI) ions.

1st transition series: manganese

For A-level you focus on manganese compounds in which the element shows a wide range of oxidation states. The +2, +4 and +7 states are most important but you may also meet the +6 state.

Manganese(VII) compounds

The jargon

$KMnO_4$ is still commonly known as potassium permanganate.

→ For A-level the most important compound of manganese is potassium manganate(VII), $KMnO_4$.
→ Potassium manganate(VII) is used as an oxidizing agent in titrimetric (volumetric) analysis.

We cannot prepare directly (by dissolving $KMnO_4(s)$ crystals in water) solutions of precise concentration and they deteriorate on standing especially in the presence of dust and other organic material. So

Links

See page 73: potassium manganate(VII) in volumetric analysis.

→ $MnO_4^-(aq)$ is not used as a primary standard.

We prepare chlorine in the laboratory fume cupboard by putting drops of concentrated hydrochloric acid onto $KMnO_4(s)$ crystals.

$$2KMnO_4 + 16HCl \rightarrow 2KCl + 2MnCl_2 + 8H_2O + 5Cl_2$$

Manganese(IV) compounds

→ Manganese(IV) oxide, MnO_2, occurs naturally as the mineral *pyrolusite* and is the only common manganese(IV) compound.

The black oxide is formed when manganese(II) nitrate decomposes on heating or when $MnO_4^{2-}(aq)$ disproportionates.

Don't forget!

Potassium manganate(VII) is used in organic chemistry as an oxidant and test reagent. It converts the side chains of aromatic compounds to carboxyl groups. Alkaline $KMnO_4(aq)$ loses its purple colour when it oxidizes the >C=C< bonds in alkenes.

→ Hydrated manganese(IV) oxide (brown) may form if insufficient acid is present when aqueous manganate(VII) is reduced:
$MnO_4^-(aq) + 4H^+(aq) + 3e^- \rightarrow MnO_2(s) + 2H_2O(l)$
+7   −3   +4

Links

See page 150: Reactions of alkenes.

→ Manganese(IV) oxide oxidizes concentrated hydrochloric acid to chlorine: $MnO_2(s) + 4HCl(aq) \rightarrow Cl_2(g) + MnCl_2(aq) + 2H_2O(l)$.
→ $MnO_2(s)$ catalyzes the decomposition of aqueous hydrogen peroxide: $2H_2O_2(aq) \rightarrow 2H_2O(l) + O_2(g)$.

Watch out!

If you are titrating acidified $Fe^{2+}(aq)$ with aqueous manganate(VII) and the solution starts to turn brown, add more $H_2SO_4(aq)$ to your conical flask.

Manganese(II) compounds

→ +2 is the most stable oxidation state of manganese.
→ Electronic configuration of Mn^{2+} is $1s^22s^22p^63s^23p^63d^5$.

Manganese(II) oxide
→ basic oxide
→ made by heating manganese(II) ethanedioate (compare FeO):
$MnC_2O_4(s) \rightarrow MnO(s) + CO(g) + CO_2(g)$

Watch out!

Heat $FeC_2O_4(s) \rightarrow FeO(s)$ = unstable black powder which spontaneously burns in air to higher oxides!

The reducing atmosphere of carbon monoxide prevents oxidation. The colour of the manganese(II) oxide is greyish green.

Manganese(II) hydroxide

→ Formed as a white precipitate when NaOH(aq) is added to aqueous manganese(II) sulphate:

$$Mn^{2+}(aq) + 2OH^-(aq) \rightarrow Mn(OH)_2(s)$$

The precipitate does not dissolve in excess alkali.

→ White $Mn(OH)_2(s)$ precipitate gradually turns brown as it oxidizes to manganese(III) oxide.

Manganese(II) sulphate

→ Hydrated manganese(II) sulphate crystals have a faint pink colour.

This salt forms a variety of crystalline hydrates. $MnSO_4 \cdot H_2O$, $MnSO_4 \cdot 4H_2O$ and $MnSO_4 \cdot 7H_2O$ are all known.

Reactions of the aqueous manganese(II) ion

You could investigate the reactions of aqueous manganese(II) ions using aqueous manganese(II) sulphate.

	Reagent	Result
1	NaOH(aq) or NH₃(aq)	white precipitate formed at first, Mn(OH)₂(s), gradually darkens to manganese(III) oxide
2	Na₃PO₄(aq) and NH₃(aq)	pink precipitate of manganese(II) ammonium phosphate
3	PbO₂(s) and conc. HNO₃(aq)	pink/purple colour of manganate(VII) ions produced by oxidation of Mn²⁺
4	NaBiO₃(s) stirred into cold HNO₃(aq) containing Mn²⁺(aq)	pink/purple colour of manganate(VII) ions observed on standing or filtering

Equations for tests 3 and 4:

$$5PbO_2(s) + 2Mn^{2+}(aq) + 4H^+(aq)$$
$$\rightarrow 2MnO_4^-(aq) + 5Pb^{2+}(aq) + 2H_2O(l)$$
$$5NaBiO_3(s) + 2Mn^{2+}(aq) + 14H^+(aq)$$
$$\rightarrow 2MnO_4^-(aq) + 5Bi^{3+}(aq) + 5Na^+(aq) + 7H_2O(l)$$

→ Lead(IV) oxide in nitric acid and sodium bismuthate(V) in nitric acid can oxidize manganese(II) to manganate(VII).

Checkpoint 1

Write the formula for manganese(III) oxide and suggest an equation for its formation from Mn(OH)₂.

Action point

Make a list of hydroxides precipitated by NaOH(aq) from aqueous transition metal ions and note whether or not they dissolve in excess alkali.

Checkpoint 2

Outline how you would prepare a sample of manganese(II) carbonate from manganese(II) sulphate.

The jargon

Manganese(II) ammonium phosphate = NH₄MnPO₄·H₂O(s).

Watch out!

Sodium bismuthate(V) is an extremely sensitive test for Mn²⁺(aq). You need to stir only a little of the solid into the acidified (with HNO₃) solution you are testing.

Exam practice (7 minutes) answers: page 140

(a) How many moles of each of the following would be needed to reduce one mole of manganate(VII) ions to manganese(II) ions in excess acid?

(i) iron(II) ions

(ii) ethanedioate ions

(iii) hydrogen peroxide molecules

(b) (i) Write a balanced ionic equation for the oxidation of SO_3^{2-}(aq) to SO_4^{2-}(aq) by manganate(VII) ions which are reduced to manganese(II) ions in excess acid. (ii) 20.0 cm³ of a solution containing SO_3^{2-}(aq) ions react with 16.0 cm³ of aqueous potassium manganate(VI) of concentration 0.015 mol dm⁻³, in acidic solution. Calculate the concentration of SO_3^{2-}(aq) ions in mol dm⁻³.

1st transition series: copper

We can tell copper from all other metals by its colour. The blue colour of hydrated copper(II) sulphate crystals and solution is another distinguishing feature of this element.

The element

The standard electrode potential for Cu^{2+}/Cu is +0.34 volt. For the other nine metals in the first transition series from Sc to Zn the value of $E^{\ominus}$ is negative, so copper is the least reactive of these ten elements.

→ Copper does not react with acids like HCl(aq) to give hydrogen.
→ Copper reacts with oxidizing acids to form copper(II) compounds and various acid reduction products depending on the conditions
   1 $Cu(s) + 2H_2SO_4(\text{hot conc.}) \rightarrow CuSO_4(aq) + 2H_2O(l) + SO_2(g)$
   2 $3Cu(s) + 8HNO_3(\text{dil.}) \rightarrow 3Cu(NO_3)_2(aq) + 4H_2O(l) + 2NO(g)$
   3 $Cu(s) + 4HNO_3(\text{conc.}) \rightarrow Cu(NO_3)_2(aq) + 2H_2O(l) + 2NO_2(g)$

When copper metal is exposed to the atmosphere, it gradually becomes covered with a greenish-blue layer (*verdigris*). The layer is usually a basic form of copper(II) carbonate, chloride or sulphate but its composition varies from place to place, e.g. by the sea it is mainly copper(II) chloride. When heated in air or oxygen copper becomes coated with black copper(II) oxide: $2Cu(s) + O_2(g) \rightarrow 2CuO(s)$.

Copper(II) compounds

Copper(II) oxide, CuO

→ Basic oxide prepared in the laboratory by heating copper(II) carbonate or nitrate: $2Cu(NO_3)_2 \rightarrow 2CuO + 2NO_2 + O_2$

Copper(II) hydroxide, Cu(OH)$_2$

→ Precipitated from aqueous copper(II) salts by aqueous ammonia or NaOH(aq): $Cu^{2+}(aq) + 2OH^-(aq) \rightarrow Cu(OH)_2(s)$

Its slight solubility in *excess concentrated* sodium hydroxide indicates that this basic oxide has some amphoteric character.

Copper(II) carbonate

→ Precipitated from aqueous copper(II) sulphate by $Na_2CO_3(aq)$ (highly alkaline) as a *basic carbonate*, e.g. $2CuCO_3 \cdot Cu(OH)_2(s)$

Basic carbonates vary in composition and occur naturally in ores, e.g. *azurite* $2CuCO_3 \cdot Cu(OH)_2$ and *malachite* $CuCO_3 \cdot Cu(OH)_2$.

Aqueous copper(II) cation

Copper(II) oxide, hydroxide or carbonate react with *excess* $H_2SO_4(aq)$ to give a blue (acidic) solution from which we can obtain blue crystals of copper(II) sulphate pentahydrate, $CuSO_4 \cdot 5H_2O$. Four water ligands are attached to each copper ion; the other molecule of water is *hydrogen bonded* within the lattice.

→ The complex ion $[Cu(H_2O)_6]^{2+}$ has a distorted octahedral structure with two water molecules at its apices and the other four ligands

---

Grade booster

You should know the names of SO$_2$, NO, NO$_2$ and N$_2$O$_4$. In equation 3 N$_2$O$_4$ would be accepted instead of 2NO$_2$.

Watch out!

Copper does not give hydrogen with dilute hydrochloric and dilute sulphuric acids. It will react with these acids in the presence of an oxidizing agent such as hydrogen peroxide.

Grade booster

Top grade candidates would know that copper exposed to the atmosphere acquires a greenish-blue layer of basic copper carbonate or, by the sea, copper chloride called verdigris. When you heat copper in air it forms a black coating of copper(II) oxide.

The jargon

Latin *verdigris* = green of Greece.

The jargon

CuSO$_4$·5H$_2$O is also called copper(II) sulphate-5-water.

(square) coplanar with the central cation.

→ Aqueous copper(II) ions take part in ligand exchange reactions where one or more H_2O molecules are replaced by stronger nucleophiles like NH_3 and EDTA.

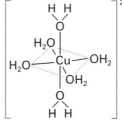

Watch out!

Some textbooks still give the formula of the aqueous copper(II) ion as $[Cu(H_2O)_4]^{2+}$.

When you add drops of aqueous ammonia to copper(II) salt solutions, you can expect to see a pale blue precipitate (copper(II) hydroxide) form and then dissolve in excess ammonia to give a deep blue solution containing tetraamminecopper (II) ions.

$$[Cu(H_2O)_6]^{2+}(aq) + 4NH_3(aq) \rightleftharpoons [Cu(H_2O)_2(NH_3)_4]^{2+}(aq) + 4H_2O(l)$$

When you add drops of concentrated hydrochloric acid to aqueous copper(II) sulphate, you can expect the solution to change from blue to green.

$$[Cu(H_2O)_6]^{2+}(aq) + 4Cl^-(aq) \rightleftharpoons [CuCl_4]^{2-}(aq) + 6H_2O(l)$$
$$\text{blue} \qquad\qquad\qquad \text{yellow/brown}$$

Don't forget!

In liquid ammonia it is possible to form copper(II) ions with six ammonia ligands, i.e. $[Cu(NH_3)_6]^{2+}$.

If you dilute the green solution with water, the blue colour returns.

→ The tetrachlorocuprate(II) ion, $[CuCl_4]^{2-}(aq)$, has a tetrahedral shape.

Copper(I) compounds

Copper(I) oxide

→ Cu_2O forms as a red solid when aldehydes and sugars reduce Fehling's solution (a *complex* Cu^{2+} ion in alkaline conditions).
The solid reacts with acids to form copper metal and copper(II) salts:
$Cu_2O(s) + 2H^+(aq) \rightarrow Cu(s) + Cu^{2+}(aq) + H_2O(l)$ (disproportionation)

Watch out!

In Fehling's solution Cu^{2+} is complexed with a bidentate ligand to stop $Cu(OH)_2$ precipitating in the alkaline conditions. Top grade candidates know that not all sugars are reducing sugars!

Copper(I) halides

→ CuCl is prepared as a white solid by reducing copper(II) chloride in hot concentrated HCl with copper and pouring the mixture into water: $Cu(s) + CuCl_2(conc \cdot HCl) \rightarrow 2CuCl(conc \cdot HCl) \rightarrow 2CuCl(s)$
→ CuBr is prepared in a similar way but copper(I) iodide is formed when aqueous copper(II) ions oxidize aqueous iodide ions:
$2Cu^{2+}(aq) + 4I^-(aq) \rightarrow 2CuI(s) + I_2(aq)$

The iodine may be measured by titrating with aqueous sodium thiosulphate using starch indicator.

→ Aqueous copper(I) ions spontaneously disproportionate unless stabilized as a complex, e.g. $[CuCl_2]^-(aq)$ and $[CuCl_4]^{3-}(aq)$

Action point

Look up the practical details (including the colour change of the indicator and the equation for the reaction) of iodine–thiosulphate titrations.

Exam practice (8 minutes) answers: page 140

(a) Describe and explain what happens to aqueous copper(II) chloride on adding (i) iron filings, (ii) aqueous ammonia until in excess, (iii) sulphur dioxide.

(b) Calculate the volume of 0.100 mol dm^{-3} HCl(aq) which will react with 0.05 mol of basic copper(II) carbonate: $2CuCO_3 \cdot Cu(OH)_2$.

Structured exam question

answers: page 141

This question concerns patterns in the properties of the hydrides, oxides and chlorides of elements shown on the right in a portion of the periodic table.

Li	Be	B	C				
Na	Mg	Al	Si	P	S	Cl	Ar
			Ge				
			Sn				
			Pb				

(a) Using only those elements listed above,

   (i)   give the formula of an ionic hydride:

....

....

   (ii) give the formula of a diatomic covalent hydride and explain why it is polar:

....

....

   (iii) give the formula of a non-polar covalent hydride and explain why it is non-polar:

....

....

   (iv) give the equation for the reaction of a diatomic hydride with water to form an alkaline solution:

....

....

(b) State how the acid–base character of the oxides changes across the period from sodium to chlorine and illustrate your statement by writing an equation for an appropriate reaction for each of the following oxides: statement:

....

....

sodium oxide:

....

sulphur dioxide:

....

(c) Describe a *trend* in the properties, *down the group* from carbon to lead, of

   (i)   the oxides CO_2, SiO_2, SnO_2, PbO_2:

....

....

   (ii) the chlorides CCl_4, $SiCl_4$, $SnCl_4$, $PbCl_4$:

....

....

(d) Compare the reactions of sodium chloride, aluminium chloride and silicon chloride with water to illustrate how the ease of hydrolysis varies across the period Na to Ar.

....

....

(e) Write an equation for the action of heat upon (i) lithium carbonate, (ii) magnesium carbonate and suggest why the other group I carbonates do not readily decompose.

....

....

(20 min)

Answers
Inorganic chemistry

Patterns and trends in the periodic table

Checkpoints

1 Noble (inert) gases, group 0
2 (a) $MgO + 2H^+ \rightarrow Mg^{2+} + H_2O$
  (b) $CO_2 + OH^- \rightarrow HCO_3^-$
  (c) (i) $Al_2O_3 + 6H^+ \rightarrow 2Al^{3+} + 3H_2O$
      (ii) $Al_2O_3 + 2OH^- + 3H_2O \rightarrow 2[Al(OH)_4]^-$

Examiner's secrets

> You will not lose marks if you omit state symbols unless we specifically tell you to include them in an equation.

Exam practice

(a) (i) Aluminium oxide
   (ii) Germanium
   (iii) Lead(IV) oxide
(b) Oxides become less acidic with increasing atomic number.
The +2 oxidation state becomes more stable
All valid answers score marks when there is more than one possible answer.

Groups I and II: alkali and alkaline earth metals

Checkpoints

1 (a) $Mg + H_2O \xrightarrow{steam} MgO + H_2$
  (b) $Ba + 2H_2O \rightarrow Ba(OH)_2 + H_2$
2 (a) (i) Hard water is water that does not readily lather with soap since it contains magnesium and/or calcium ions which react with soaps to form an insoluble scum.
    (ii) Limescale is calcium carbonate formed by the decomposition of hydrogencarbonate ions in temporarily hard water.
    $$Ca^{2+} + 2HCO_3^- \rightarrow CaCO_3 + CO_2 + H_2O$$
  (b) Stalactites and stalagmites form when water evaporates – solid $Ca(HCO_2)_2$ is too unstable to exist:
    $$Ca^{2+} + 2HCO_3^- \rightarrow CaCO_3 + CO_2 + H_2O$$
  (c) Small highly charged ions distort the symmetry of the carbonate ion so that it decomposes to an oxide ion and carbon dioxide. Although the calcium ion is doubly charged the smaller lithium ion has the greater polarizing power and lithium carbonate decomposes at a lower temperature than calcium carbonate.

Exam practice

(a) (i) $CaO + H_2O \rightarrow Ca(OH)_2$
   (ii) $BaO_2 + 2H_2O \rightarrow Ba(OH)_2 + H_2O_2 \rightarrow H_2O + \frac{1}{2}O_2$ [in hot water]
   (iii) $LiH + H_2O \rightarrow LiOH + H_2$
(b) (i) $3Mg + N_2 \rightarrow Mg_3N_2$
   (ii) $TiCl_4 + 2Mg \rightarrow 2MgCl_2 + Ti$
   (iii) $NO_3^- + 4Mg + 10H^+ \rightarrow 4Mg^{2+} + 3H_2O + NH_4^+$

Industrial chemistry: s-block

Checkpoints

1 (a) (i) Anode (ii) Cathode
  (b) This is to separate the products.
  (c) Membrane cells are more efficient and environmentally friendly. In the mercury cathode cell, small amounts of toxic mercury escape into the environment and in the diaphragm cells some chlorine is lost.
2 (a) $2NaCl + CaCO_3 \rightarrow CaCl_2 + Na_2CO_3$
  (b) (i) Ammonia is very volatile and very soluble; therefore there will always be some losses.
     (ii) Changes in industrial practice are always made on the basis of economics. The use of a carbonate mineral may also present fewer environmental problems which is reflected in costs.
3 (a) Chlorine corrodes steel.
  (b) To separate the products of electrolysis.
  (c) Some calcium may be discharged at the cathode which will alloy with the sodium.
4 (a) $CaCO_3 \rightarrow CaO + CO_2$
  (b) (i) A reducing agent
     (ii) Calcium   +2 to zero
         Aluminium   zero to +3

Exam practice

(a) (i) When magnesium is heated in air it burns with a bright white flame, forming both the oxide and nitride:
    $Mg + \frac{1}{2}O_2 \rightarrow MgO$
    $3Mg + N_2 \rightarrow Mg_3N_2$
   (ii) A reaction occurs but only the nitride reacts:
    $$Mg_3N_2 + 6H_2O \rightarrow 3Mg(OH)_2 + 2NH_3$$
   (iii) The red litmus turns blue.
(b) (i) Magnesium carbonate and lithium carbonate both decompose on heating.
   (ii) Beryllium oxide and aluminium oxide are both amphoteric.

Group III: aluminium and boron

Checkpoints

1 Boron $1s^2 2s^2 2p^1$
  Aluminium $1s^2 2s^2 2p^6 3s^2 3p^1$
2 (a) Ammonia will be pyramidal (three bonding pairs and one non-bonding pair of electrons).
  (b) $H_3N—BCl_3$ will be tetrahedral around both the nitrogen atom and the boron atom as each is surrounded by four bonding pairs of electrons.
  (c) The tetrahydridoborate(III) ion is tetrahedral as there are four bonding pairs of electrons in the boron valence shell.
3 The white precipitate is aluminium hydroxide and the colourless gas is carbon dioxide. Sodium carbonate (the salt of a strong base and weak acid) is hydrolyzed and so has an alkaline solution. The aqueous aluminium ion is acidic.

$$2Al^{3+} + 3Na_2CO_3 + 3H_2O \rightarrow 2Al(OH)_3 + 3CO_2 + 6Na^+$$

Exam practice

When aqueous ammonia is added to aqueous aluminium sulphate, a white precipitate is formed which is insoluble in excess of the reagent:

$$Al^{3+}(aq) + 3OH^-(aq) \rightarrow Al(OH)_3(s)$$

When aqueous ammonia is added to aqueous zinc sulphate, a white precipitate is formed which is soluble in excess aqueous ammonia to form a colourless solution:

$$Zn^{2+}(aq) + 2OH^-(aq) \rightarrow Zn(OH)_2(s)$$
$$Zn(OH)_2(s) + 4NH_3(aq) \rightarrow [Zn(NH_3)_4]^{2+} + 2OH^-(aq)$$

Group IV: elements and oxides

Checkpoints

1 Metals are ductile, malleable and sonorous.
2 (a) A, Diamond; B, graphite; C, fullerene, C_{60}.
2 (b) A rolled-up graphite-like layer structure with a diameter about ten thousand times smaller than a human hair.
3 (a) Iron is more reactive than tin so the iron corrodes (rusts) when the metals are exposed to moist air:
$$Fe \rightarrow Fe^{2+} + 2e^-$$
Zinc is more reactive than iron so the zinc is oxidized when the metals are exposed to moist air:
$$Zn \rightarrow Zn^{2+} + 2e^-$$
These electrons suppress the oxidation of iron.
  (b) Lead was used in the production of leaded ('antiknock') petrol. Phasing out this fuel caused a decrease in demand for lead.
4 (a) Silicon dioxide and silicates are refractory materials (unaffected by very high temperatures).
  (b) The atoms in silicon dioxide are held firmly in the giant (tetrahedral) lattice structure by strong covalent bonds (compare diamond). Thus the compound acts as an abrasive by removing atoms from other structures without losing atoms from its own structure.

Exam practice

(a) When carbon dioxide is passed into limewater, a white precipitate of calcium carbonate is observed:
$$Ca(OH)_2 + CO_2 \rightarrow CaCO_3 + H_2O$$
Because calcium hydroxide is only sparingly soluble, the carbon dioxide quickly removes all the calcium and hydroxide ions. Excess carbon dioxide then reacts with carbonate ions to form aqueous hydrogencarbonate ions and the white precipitate dissolves to give a colourless solution: $CO_2 + CO_3^{2-} + H_2O \rightarrow 2HCO_3^-$
(b) In the carbonate ion, the central carbon is bonded to three equivalent oxygen atoms. VSEPR theory predicts a trigonal planar ion with more than one arrangement of bonds. So the structure has delocalized electrons (dotted lines).

Group IV: chlorides and hydrides

Checkpoints

1 (a) Lead(II) nitrate(V); (b) plumbate(II) ion.
2 (a) You might think it is energetically unfavourable (see pages 81–85) but $\Delta H^\ominus = -139$ kJ mol for liquid tetrachloromethane. So it must be kinetically unfavourable with an activation energy so high that the reaction does not take place at an appreciable rate.
  (b) Aqueous iodine is only a very dilute solution of I_2 (0.0013 mol dm^{-3} at 25 °C) and just perceptibly yellow/brown. The covalent iodine molecule is very much more soluble in covalent organic solvents. When shaken with tetrachloromethane (which is immiscible with water and more dense), almost all the iodine dissolves in the CCl_4 giving a violet colour and leaving the aqueous layer colourless.
3 Assuming the HCl is gaseous, then

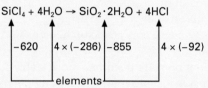

$$SiCl_4 + 4H_2O \rightarrow SiO_2 \cdot 2H_2O + 4HCl$$

Hence $\Delta H = -855 - 368 + 620 + 1144 = +541$ kJ mol^{-1}
4 Hexachlorostannate(IV) ion; hexachloroplumbate(IV) ion

Exam practice

(a) (i) $Si(s) + 2Cl_2 \rightarrow SiCl_4(l)$ Heat silicon powder in dry chlorine gas and collect product in a dry receiver.
  (ii) $Pb(s) \rightarrow Pb^{2+}(aq) + 2Cl^-(aq) \rightarrow PbCl_2(s)$ Dissolve lead powder in nitric acid in a fume cupboard, add sodium chloride to the aqueous lead(II) nitrate and filter off the white precipitate.
(b) (i) The chloride is violently hydrolyzed to SiO_2 and NaCl.
  (ii) The chloride dissolves to form a solution containing the tetrachloroplumbate(II) anion $[PbCl_4]^{2+}$.

Group V: elements and oxides

Checkpoints

1 Oxygen and the noble gases, mainly argon but all the others as well.
2 Covalent radius decreases L → R across a period so a boron atom is bigger than a nitrogen atom. Hence ⚘ represents boron.
3 Glowing splint thermally decomposes dinitrogen oxide into a mixture of 33% oxygen 67% nitrogen. This concentration of oxygen is enough to re-light the splint.

Exam practice

(a) (i) $2NO + O_2 \rightarrow 2NO_2$. Nitrogen dioxide is a brown gas
  (ii) $2N_2O \rightarrow 2N_2 + O_2$. Glowing splint thermally decomposes dinitrogen oxide into a mixture of 33% oxygen 67% nitrogen. This concentration of oxygen is enough to re-light the splint.
(b) (iii) Sodium nitrite $NaNO_2$,
  (ii) Sodium nitrate $NaNO_3$,

(iii) Sodium phosphate(III) Na_3PO_3,

(iv) Sodium phosphate(V) Na_3PO_4.

Group V: oxoacids and hydrides

Checkpoints

1 $N_2O_4 + 2NaOH \rightarrow NaNO_3 + NaNO_2 + H_2O$

2 (a) pH = $-\log_{10}(0.02)$ = 1.7

(b) Use a titrimetric method to find the volume of nitric acid which will exactly neutralize a given volume of aqueous sodium hydroxide. Repeat the experiment using exactly the same volumes of acid and alkali but omit the indicator. Evaporate the resulting solution to small bulk and allow the salt to crystallize.

3 $H_2PO_4^-$

4 $PCl_5 + H_2O \rightarrow POCl_3 + 2HCl$

5 (a) When heated ammonium chloride undergoes thermal dissociation:

$$NH_4Cl(s) \rightleftharpoons NH_3(g) + HCl(g)$$

On the cooler sides of the tube the products recombine to form the white solid ammonium chloride.

(b) However, the ammonia molecule is lighter than the hydrogen chloride molecule and so diffuses faster. Ammonia gas reaches the end of the test tube before hydrogen chloride gas and the alkaline gas turns moist red litmus paper blue:

$$NH_3(g) + H_2O(l) \rightleftharpoons NH_4^+(aq) + OH^-(aq)$$

Exam practice

(a) $N_2O_4 + 2NaOH \rightarrow NaNO_2 + NaNO_3 + H_2O$. Oxidn nos. +3 and +5

(b) (i) 0.1 mol PCl_3 reacts with 0.3 mol H_2O.
Molar mass H_2O is 18.0 g mol^{-1}. So mass water needed is 0.3 mol × 18.0 g mol^{-1} = 5.4 g.

(ii) 0.1 mol PCl_3 could give 0.3 mol HCl or 0.3 × 24 dm^3 = 7.2 dm^3 at room temperature and pressure.

Group VI: oxygen and sulphur

Checkpoint

$KNO_3 \rightarrow KNO_2 + \frac{1}{2}O_2$

$HgO \rightarrow Hg + \frac{1}{2}O_2$

$2Pb_3O_4 \rightarrow 6PbO + O_2$

You would get the marks for any other correct examples.

Exam practice

(a) (i) When sulphur is heated it becomes a mobile amber liquid as the S_8 rings break away from their positions in the solid sulphur lattice and become free to move in the liquid state. As the temperature rises S_8 rings break open and the liquid darkens because of unpaired electrons. The viscosity of the liquid increases as broken rings combine into longer sulphur atom chains which interact with one another. The decrease in the mobility of the liquid reaches a maximum (around 180 °C). As the temperature rises to the boiling point of 444 °C, the long chains begin to break up and the liquid becomes more mobile (smaller molecules) and much darker (more unpaired electrons).

(ii) When the sulphur is poured into cold water, the chains of sulphur atoms persist, giving the rubber-like form known as plastic sulphur which only slowly reverts to the S_8 structure of rhombic sulphur.

(b) (i) $Mg + S \rightarrow MgS$ (ii) $Cl_2 + 2S \rightarrow S_2Cl_2$

(c) Sulphur has an oxidation state of –2 in the sulphide ion since it is far more electronegative than sodium. In covalent sulphur hexafluoride, fluorine is the more electronegative and forces sulphur into an oxidation state of +6.

Group VI: water and hydrogen peroxide

Checkpoints

1 (a) H_2O

(b) O^{2-}

In $H_2O(l) + H_2O(l) \rightleftharpoons H_3O^+(aq) + OH^-(aq)$
water acts as both a Brønsted–Lowry acid (proton donor) and a Brønsted–Lowry base (proton acceptor).

2 An ion-exchange resin softens hard water by removing calcium ions and replacing them with sodium ions:
$[resin(2Na^+)_n] + Ca^{2+}(aq) \rightarrow [resin(Ca^{2+})_n] + 2Na^+(aq)$
Eventually the resin runs out of sodium ions. Salt provides sodium ions to regenerate the resin by displacing the calcium ions it removed from the water:
$[resin(Ca^{2+})_n] + 2Na^+(aq) \rightarrow [resin(2Na^+)_n] + Ca^{2+}(aq)$

Exam practice

The two ion–electron half-equations are
$$MnO_4^- + 8H^+ + 5e^- \rightarrow Mn^{2+} + 4H_2O$$
$$2H_2O_2 \rightarrow 2H_2O + O_2 + 2e^-$$
Therefore 2 mol $MnO_4^- \equiv$ 5 mol H_2O_2
No. of mol of MnO_4^- = $(37.5 \times 10^{-3} \times 0.0187)$
$$= 7.01 \times 10^{-4} \text{ mol}$$
No. of mol H_2O_2 in 25.0 cm^3 of the solution
$$= 2.5 \times 7.01 \times 10^{-4} \text{ mol}$$
No. of mol H_2O_2 in 1.00 dm^3
$$= 2.5 \times 7.01 \times 10^{-4} \text{ mol} \times \frac{1\,000}{25.0}$$
$$= 0.070\,1 \text{ mol}$$
Concentration of the aqueous hydrogen peroxide = 0.0701 mol dm^{-3}

Group VII: halogens and hydrogen halides

Checkpoints

1 (a) In the reaction with water $Cl_2 + H_2O \rightarrow HClO + HCl$ the oxidation state of chlorine changes from zero to +1 *and* –1. Disproportionation is the simultaneous oxidation and reduction of the same element.

(b) All covalent non-metal chlorides formed by direct

combination of the element with chlorine are hydrolyzed by water and many metal chlorides, e.g. aluminium chloride and iron(III) chloride, formed by direct combination are also hydrolyzed. So we must use anhydrous reagents and keep apparatus dry.

2 (a) $NH_3 + HCl \rightarrow NH_4Cl$
   (b) $2KMnO_4 + 16HCl \rightarrow 2MnCl_2 + 2KCl + 5Cl_2 + 8H_2O$

Exam practice

(a) (i) $2NaOH + Cl_2 \rightarrow NaCl + NaClO + H_2O$ −1 and +1
   (ii) $6NaOH + 3Cl_2 \rightarrow 5NaCl + NaClO_3 + 3H_2O$ −1 and +5
(b) The HF_2^- is linear (two bonding pairs of electrons and no non-bonding pairs).

Group VII: halides and interhalogen compounds

Checkpoints

1 When sulphuric acid reacts with HBr the change in oxidation state of sulphur is +6 to +4; with HI the change is +6 to −2. Therefore, HI is the stronger reducing agent.
2 $NaHCO_3 + HCl \rightarrow NaCl + H_2O + CO_2$
3 (a) (i) In alkaline solution, silver oxide, Ag_2O, is precipitated.
       (ii) Nitric acid must be used since hydrochloric would produce a precipitate of silver chloride, and sulphuric acid, a precipitate of silver sulphate.
   (b) The silver halides become increasingly covalent in character.
   (c) The lattice enthalpy of calcium fluoride is large enough for it to be insoluble.
4 (a) Iodine chloride
   (b) Iodine trichloride

Exam practice

(a) (i) $2Al + 3Cl_2 \rightarrow Al_2Cl_6$
   (ii) $2MnO_4^- + 16H^+ + 10I^- \rightarrow 5I_2 + 2Mn^{2+} + 8H_2O$
(b) One mole of iodate(V) ions liberates three moles of iodine. Therefore 0.15 mol I_2 is liberated by 0.05 mol IO_3^-.

$$\text{Concentration} = \left(\frac{1\,000}{20}\right) \times 0.05\ \text{mol dm}^{-3} = 2.5\ \text{mol dm}^{-3}$$

Industrial chemistry: aluminium and carbon

Checkpoint

Compound readily hydrolyzed because the central aluminium atom (unlike boron in $NaBH_4$) has d-orbitals available to accept a lone pair of electrons from a water molecule. (Compare with water on CCl_4 and $SiCl_4$.)

Exam practice

(a) (i) Filtration. (ii) Inefficient electrolysis and impure product. (iii) Some cryolite is lost because the reactions in the cell are complex and involve fluoride ions.
(b) Aqueous sulphuric acid is electrolyzed and oxygen

liberated at the aluminium anode reacts with the metal surface to form aluminium oxide. Anodizing thickens the oxide layer on aluminium surfaces to adsorb dyes and permanently colour the metal surface.

Industrial chemistry: silicon and nitrogen

Checkpoints

1 The oxidation state is +4.
2 Asbestos dust can enter the body and damage the lungs. Blue asbestos is very dangerous and listed as a carcinogen causing asbestosis, a disease which develops to produce a variety of carcinoma.
3 Unconverted hydrogen and nitrogen are recycled.

Exam practice

(a) Process converts atmospheric nitrogen into a form in which it can be assimilated by plants.
   Those conditions of temperature and pressure most favourable to producing the most economic yield of ammonia.
   A catalyst in a *different* phase from the reactants.
(b)
$$\frac{p^2_{NH_3}}{p_{N_2} \times p^3_{H_2}} = K_p$$
   (i) If at a given temperature (K_p is constant) the pressure is increased, then the denominator would momentarily increase. To bring K_p back to its constant value the partial pressure of NH_3 must increase. Therefore an increase in pressure favours the formation of ammonia.
   (ii) Cooling will favour formation of ammonia – reaction is exothermic so K_p increases as temperature falls. Removing ammonia encourages more nitrogen and hydrogen to combine trying to reach equilibrium.
(c) (i) Convenient, economic and likely to produce a finely divided form of iron with a large surface area.
   (ii) $Fe_2O_3 + 3H_2 \rightarrow 2Fe + 3H_2O$

Industrial chemistry: sulphur and the halogens

Checkpoints

1 (a) (i) Impurities like arsenic oxide poison the catalyst and make it less efficient.
       (ii) The percentage conversion would decrease.
   (b) (i) A catalyst which is in a different phase from the reactants.
       (ii) A promoter is a substance which when added to a catalyst improves the catalytic effect.
2
$$\frac{p^2_{SO_3}}{p^2_{SO_2} \times p_{O_2}} = K_p$$
The expression for K_p from the law of chemical equilibrium shows that its units are (pressure)$^{-1}$.
If at a given temperature (K_p is constant) the pressure is increased, then the denominator would momentarily increase. To bring K_p back to its constant value the partial

pressure of SO_3 must increase. Therefore an increase in pressure favours the formation of sulphur trioxide.

As the forward reaction is exothermic, an increase in temperature decreases the value of K_p and therefore the yield of sulphur trioxide decreases.

3 CFCs are chlorofluorocarbons – very stable compounds which have been used as aerosol propellants and refrigerants. They can persist in the atmosphere, reach the stratosphere and deplete the ozone layer by taking part in free radical reactions. The EU has banned their manufacture in order to allow the ozone layer to recover.

Exam practice

(a) Sulphur dioxide and excess air are purified and passed over a heterogeneous catalyst of promoted vanadium(V) oxide at an optimum temperature of about 430 °C:

$$2SO_2(g) + O_2(g) \rightleftharpoons 2SO_3(g)$$

Sulphur trioxide gas is absorbed in 98% sulphuric acid to form oleum ($H_2S_2O_7$). The oleum is diluted to form 100% sulphuric acid (H_2SO_4).

(b) (i) Colourless steamy fumes evolve. Volatile hydrogen chloride displaced by less volatile sulphuric acid: $H_2SO_4 + NaCl \rightarrow NaHSO_4 + HCl$.

(ii) Brown fumes condensing to drops of pale yellow fuming liquid. Volatile nitric acid displaced by less volatile sulphuric acid: $H_2SO_4 + KNO_3 \rightarrow KHSO_4 + HNO_3$. Colour caused by thermal decomposition of nitric acid into brown nitrogen dioxide:

$$2HNO_3 \rightarrow 2NO_2 + H_2O + \tfrac{1}{2}O_2.$$

(iii) Purple vapour condensing to shiny black crystals, some steamy fumes and smell of bad eggs. Sulphuric acid displaces hydrogen iodide which thermally decomposes and also reduces sulphuric acid to hydrogen sulphide:

$$H_2SO_4 + KI \rightarrow KHSO_4 + HI \cdot 2HI \rightarrow H_2 + I_2$$
$$8HI + H_2SO_4 \rightarrow H_2S + 4H_2O + 4I_2$$

(iv) Fruity pear drop smell produced. Sulphuric acid catalyzes formation of pentyl ethanoate ester:

$$CH_3CO_2H + C_5H_{11}OH \rightleftharpoons CH_3CO_2C_5H_{11} + H_2O$$

(c) (i) Sea water; (ii) sodium iodate in Chile saltpetre.

(d) Sea water is acidified and the bromide ions oxidized to bromine by chlorine in a non-metal displacement reaction: $Cl_2 + 2Br^- \rightarrow 2Cl^- + Br_2$.

Sodium iodate(V) is reduced to iodine using sulphur dioxide: $5SO_2 + 2KIO_3 + 4H_2O \rightarrow I_2 + 4H_2SO_4 + K_2SO_4$

(e) In the presence of dilute nitric acid, aqueous silver nitrate produces a white precipitate with aqueous chloride ions which is soluble in aqueous ammonia, a cream precipitate with aqueous bromide ions which is partially soluble in aqueous ammonia, a primrose yellow precipitate with aqueous iodide ions which is insoluble in aqueous ammonia, but no precipitate with aqueous fluoride ions since silver fluoride is soluble.

(f) Hydrochloric acid for cleaning iron and steel surfaces prior to galvanizing, spray-painting and tin-plating. Chloroethene used as the monomer in the production of poly(chloroethene) or PVC.

Group VIII (or 0): the noble gases

Checkpoints

1 Noble metals are very unreactive d-block metals below hydrogen in the electrochemical series.

Rh Pd Ag
Ir Pt Au

2 (a) The energy required to bring about

$$O_2(g) \rightarrow O_2^+(g) + e^-$$

(b) More energy may be needed to break the F-F bond than may be released when F-He or F-Ne bonds form. ΔG for the reaction would be positive and so the reaction would be not be energetically feasible

Exam practice

(a) Ar $1s^2 2s^2 2p^6 3s^2 3p^6$

A chlorine atom Cl $1s^2 2s^2 2p^6 3s^2 3p^5$ gains an electron to be-come a chloride anion $\underline{Cl}^-$ $1s^2 2s^2 2p^6 3s^2 3p^6$ and a potassium atom K $1s^2 2s^2 2p^6 3s^2 3p^6 4s^1$ loses an electron to become a potassium cation K^+ $1s^2 2s^2 2p^6 3s^2 3p^6$. The chloride ion, potassium ion and argon atom are isoelectronic.

(b) Pass a dry mixture of xenon and fluorine gases over hot metallic nickel, acting as a catalyst, in the absence of air and moisture.

XeF_2: Here we have three non-bonding pairs of electrons and two bonding pairs. This gives a linear structure.

XeF_4: Here there are four bonding pairs of electrons and two non-bonding pairs giving a square planar structure:

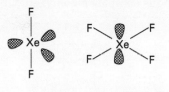

represents a non-bonding pair of electrons

1st transition series: metals, aqueous ions and redox

Checkpoints

1 (a) $TiCl_4 + 4Na \rightarrow Ti + 4NaCl$
(b) $Ni + 4CO \rightleftharpoons Ni(CO)_4$
(c) $Cu^{2+} + 2e^- \rightarrow Cu$

2 (a)

$ScCl_3$	$TiCl_4$	VCl_5	$CrCl_3$	$MnCl_2$
$FeCl_3$	$CoCl_2$	$NiCl_2$	$CuCl_2$	$ZnCl_2$

(b)

$Sc_2(SO_4)_3$	$Fe_2(SO_4)_3$
$Ti(SO_4)_2$	$CoSO_4$
$V_2(SO_4)_5$	$NiSO_4$
$Cr_2(SO_4)_3$	$CuSO_4$
$MnSO_4$	$ZnSO_4$

Exam practice

(a) The electronic configuration is Zn
$1s^2 2s^2 2p^6 3s^2 3p^6 3d^{10} 4s^2$ and the d-shell electrons show zinc to be a d-block element. The d-shell is always full, so the oxidation state of zinc in its compounds is +2 and, unlike the transition metals, does not vary.

(b) (i) Orange to green to blue as the oxidation state of chromium is reduced from +6 to +3 to +2; (ii) yellow to blue to green to violet as the oxidation state of vanadium is reduced from +5 to +4 to +3 to +2.

1st transition series: redox reactions and complex ion formation

Checkpoint

Questions on free energy do not apply to every examination board.

$$\Delta G^{\ominus} = -5 \times 96\,500 \times 0.74 \text{ J mol}^{-1} = -357\,000 \text{ J mol}^{-1}$$

Exam practice

(a) (i) A complex cation is a positively charged ion consisting of a central metal ion with molecules or ions (ligands) datively bonded to it. $[Cu(NH_3)_4]^{2+}$

   (ii) A complex anion is a negatively charged ion consisting of a central metal ion with molecules or ions (ligands) datively bonded to it. $[CuCl_4]^{2-}$

   (iii) A bidentate ligand is a molecule or anion capable of forming two dative bonds with a central metal ion. 1,2-diaminoethane $NH_2CH_2CH_2NH_2$

(b) (i) The carbonate would neutralize the citric acid, make the solution alkaline and the iron would precipitate as a basic hydroxide. The iron(II) cations would remove poisonous cyanide ions by forming a stable non-poisonous complex anion, $[Fe(CN)_6]^{4-}$

   (ii)

$$
\begin{array}{c}
\text{H} \\
| \\
\text{H–C–CO}_2\text{H} \\
| \\
\text{HO–C–CO}_2\text{H} \\
| \\
\text{H–C–CO}_2\text{H} \\
| \\
\text{H}
\end{array}
$$

The acid lowers the pH of the solution and suppresses the hydrolysis of the iron(II) sulphate. It reacts with the sodium carbonate to form carbon dioxide which makes the mixture fizzy and easier to swallow.

1st transition series: chromium

Checkpoint

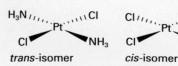

trans-isomer   *cis*-isomer

Exam practice

(a) (i) +3 because charge on the sulphate ion is 2– so charge on the chromium ion must be 3+;

(ii) +6 because oxidation state of oxygen is –2 and chlorine is –1.

(b) (i) $6NaOH(aq) + Cr_2(SO_4)_3(aq) \rightarrow$
$$3Na_2SO_4(aq) + 2Cr(OH)_3(s)$$

   (ii) $4NaCl + K_2Cr_2O_7 + 3H_2SO_4 \rightarrow$
$$2CrO_2Cl_2 + 2Na_2SO_4 + K_2SO_4 + 3H_2O$$

   (iii) $4NaOH + CrO_2Cl_2 \rightarrow$
$$Na_2CrO_4(aq) + 2NaCl(aq) + 2H_2O(l)$$

1st transition series: manganese

Checkpoints

1 $3Mn(OH)_2 \rightarrow Mn_2O_3 + 3H_2O$

2 Add a solution of sodium carbonate to aqueous manganese(II) sulphate and filter off the precipitated carbonate. Wash, dry and store.
$$MnSO_4(aq) + Na_2CO_3(aq) \rightarrow MnCO_3(s) + Na_2SO_4(aq)$$
Try writing this ordinary equation as an ionic equation.

Exam practice

(a) (i) 5 (ii) 2.5 (iii) 2.5

(b) (i) $2MnO_4^- + 6H^+ + 5SO_3^{2-} \rightarrow 2Mn^{2+} + 3H_2O + 5SO_4^{2-}$

   (ii) $0.015 \times \left(\dfrac{5}{2}\right) \times \left(\dfrac{16.0}{20.0}\right) = 0.030 \text{ mol dm}^{-3}$

1st transition series: copper

Exam practice

(a) (i) The blue colour of the solution fades and a reddish precipitate of copper forms:
$$Cu^{2+}(aq) + Fe(s) \rightarrow Fe^{2+}(aq) + Cu(s)$$

   (ii) A blue precipitate forms which dissolves in excess aqueous ammonia to form a deep blue solution:
$$Cu^{2+}(aq) + 2OH^-(aq) \rightarrow Cu(OH)_2(s)$$
With excess ammonia
$$Cu(OH)_2(s) + 4NH_3(aq) \rightarrow [Cu(NH_3)_4]^{2+}(aq) + 2OH^-(aq)$$

   (iii) With sulphur dioxide, a reducing agent, insoluble copper(I) chloride is formed as a white precipitate:
$$2CuCl_2 + SO_2 + 2H_2O \rightarrow 2CuCl + H_2SO_4 + 2HCl$$

(b) One mole of basic copper(II) carbonate produces two moles of carbonate ion and two moles of hydroxide ion.
$$2CO_3^{2-} + 4H^+ \rightarrow 2CO_2 + 2H_2O$$
$$2OH^- + 2H^+ \rightarrow 2H_2O$$
One mole of basic copper(II) carbonate requires six moles of $H^+ \equiv$ six moles of HCl.
0.05 mol basic copper(II) carbonate $\equiv$ 0.30 mol of $H^+ \equiv 3.0 \text{ dm}^3$ of HCl of concentration $0.100 \text{ mol dm}^{-3}$.

Grade booster

You might gain more marks if you gave the following equations:
$$[Cu(H_2O)_6]^{2+}(aq) + 2OH^-(aq) \rightleftharpoons [Cu(H_2O)_4(OH)_2](s) + 2H_2O(l)$$
$$[Cu(H_2O)_6]^{2+}(aq) + 4NH_3(aq) \rightleftharpoons [Cu(H_2O)_2(NH_3)_4]^{2+}(aq) + 4H_2O(l)$$
and you referred to the equilibria moving to the right as ammonia is added and the tetraamminediaquocopper(II) ion formed.

Incidentally, all six water ligands are replaced if the reaction is in liquid ammonia instead of water.

Structured Exam practice

(a) (i) Na^+H^-

(ii) H—Cl

because chlorine is more electronegative than hydrogen so the two atoms share the one pair of electrons unequally.

(iii) CH_4

because the electronegativities of carbon and hydrogen are similar and the tetrahedral shape of the molecule gives a net dipole moment of zero.

(iv) $Na^+H^-(s) + H_2O(l) \rightarrow Na^+(aq) + OH^-(aq) + H_2(g)$

Grade booster

Remember that group I metals are powerful reducers and give up electrons to form ionic compounds.

(b) Statement: oxides of Na and Mg are basic, aluminium oxide is amphoteric and the oxides of Si, P, S and Cl are acidic.

sodium oxide: $Na_2O(s) + H_2O(l) \rightarrow 2Na^+(aq) + 2OH^-(aq)$

sulphur dioxide: $SO_2(g) + H_2O(l) \rightarrow H_2SO_3(aq)$

(c) (i) The oxides change from acidic (CO_2, SiO_2) to amphoteric (SnO_2, PbO_2) and become oxidizing agents (SnO_2, PbO_2).

(ii) From CCl_4 to $PbCl_4$ the chlorides show a decrease in thermal stability and an increase in the tendency to disproportionate into the divalent state, giving off chlorine gas.

(d) Sodium chloride simply dissolves in water.

Aluminium chloride fumes in moist air dissolves exothermically in water to give an acidic solution:

$$AlCl_3(s) + 6H_2O(l) \rightarrow [Al(H_2O)_5OH]^+(aq) + H^+(aq)$$

Grade booster

You may get a bonus mark for an answer that is a bit special.

Silicon tetrachloride reacts violently with water to give dense fumes of hydrogen chloride and a gelatinous precipitate of hydrated silicon oxide (silica gel):

$$SiCl_4(l) + 4H_2O \rightarrow SiO_2 \cdot 2H_2O + 4HCl$$

(e) (i) $Li_2CO_3 \rightarrow Li_2O + CO_2$ (ii) $MgCO_3 \rightarrow MgO + CO_2$

The polarizing power of the very small lithium cation is high enough to distort and promote the decomposition of the carbonate anion. The polarizing power of the other larger group I cations is not high enough to do this.

Organic chemistry

Get used to structural formulae and you will make sense of organic chemistry. You already know H_2O (H—O—H), CO_2 (O=C=O) and therefore that hydrogen, oxygen and carbon form respectively one, two and four covalent bonds. So you could deduce the formula and structure of ethyne, the simplest hydrocarbon, as C_2H_2 (H—C≡C—H). Could you work out the structure of ethene (C_2H_4) and ethane (C_2H_6)? You must also get used to nomenclature like *eth*ane, *eth*ene, *eth*yne. How many carbon atoms in *eth*anol? You will learn about modern instruments that tell us the structure of compounds. You will also learn how we simplify our study of organic chemistry by classifying compounds, reagents (as free radicals, electrophiles and nucleophiles) and types of reaction (as addition, elimination and substitution).

By the end of this chapter you will be able to

- Give the name, structural formula and typical properties of hydrocarbons and of members of homologous series of oxygen, nitrogen and halogen containing compounds

- Describe types of isomerism, including *cis–trans* and optical isomerism, and be able to work out possible isomers for a given molecular formula

- Describe and explain the typical reactions of homologous series in terms of the structure and properties of functional groups

- Describe reactions as free radical, electrophilic addition, nucleophilic substitution and be able to write equations and state conditions for reactions

- Describe modern spectroscopic techniques and use IR, UV, NMR and mass spectra to identify and determine the structure of a particular compound

- Describe some of the principles and processes involved in the industrial production of organic chemicals and polymers

Topic checklist

Tick each of the boxes below when you are satisfied that you have mastered the topic

	Edexcel		AQA		OCR		WJEC		CCEA	
	AS	A2	AS	A2	AS	A2	AS	A2	AS	A2
How to name compounds	○		○		○		○		○	
Classes of compounds and functional groups	○		○		○		○		○	
Types of reactions and reagents	○		○		○		○		○	
Hydrocarbons: alkanes and alkenes	○		○		○		○		○	
Hydrocarbons: alkenes and arenes	○		○		○		○		○	
Compounds containing halogens	○	●	○	●	○	●	○	●	○	●
Compounds containing oxygen: alcohols, phenols, aldehydes and ketones	○	●	○	●	○	●	○	●	○	●
Compounds containing oxygen: aldehydes, ketones and carboxylic acids	○	●	○	●	○	●	○	●	○	●
Compounds containing nitrogen: amines and amino acids	○	●	○	●	○	●	○	●	○	●
Compounds containing nitrogen: proteins and polyamides		●		●		●		●		●
Instrumental techniques: UV, visible and IR spectroscopy	○	●	○	●	○	●	○	●		●
Instrumental techniques: NMR and mass spectrometry		●		●	○	●		●		●
Analytical techniques: chromatography		●		●		●		●		●
Industrial chemistry: oil refining and petrochemicals	○	●	○	●	○	●	○	●	○	●

How to name compounds

Chemists try to give organic compounds unambiguous names that tell us about the structure of the substance. We use a set of nomenclature rules to generate systematic names that may consist of prefix(es), root and suffix together with numbers and punctuation.

Naming alkanes

Many organic compounds have a root derived from the names of alkanes. Study these five rules.

→ Base the root name on the longest continuous carbon chain and the straight-chain alkane with the same number of carbons.
→ Add prefixes based on the shorter carbon branches and the names of the corresponding straight-chain alkanes.
→ Indicate the number of identical branches by adding di- (two), tri- (three), tetra- (four), etc.
→ Number the positions of the branches on the longest chain from the end giving the lower number for the initial branching point.
→ Attach the prefixes in alphabetical order of branch name.

Examples

$$CH_3CHCH_2CH_3$$
with CH₃ branch

2-methylbutane

$$CH_3CH_2CHCH—CHCH_2CH_3$$
with CH₃ CH₃ branches

3,4-dimethylheptane

Naming alkenes

We name alkenes in the same way as alkanes but with the following important differences:

→ Base the root name on the straight-chain alkane with the same number of C-atoms as the longest continuous carbon chain that contains the double bond.
→ Use a number to show the position of the double bond.
→ Use *cis*- and *trans*- as prefixes to name simple geometric isomers with the same group on each of two doubly bonded carbon atoms.

Examples

$$H_2C=CHCH_2CH_3$$
but-1-ene

$$CH_3C=C—CH_3$$
with CH₃ branch
2-methylbut-2-ene

Skeletal formulae

cis-but-2-ene

trans-but-2-ene

→ Use E- and Z- as prefixes to name *more complex* geometric isomers.

CIP rules
1. rank the two groups *at each end* of the double bond by the atomic number of their atoms, e.g., Br (35) > C (6) > H (1)

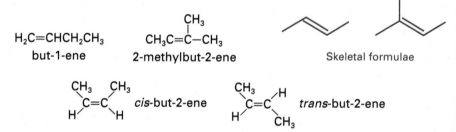

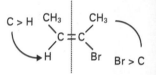

Watch out!

The first four alkanes keep their traditional names:
methane
ethane
propane
butane
Learn them by heart if you want the top grade!

The jargon

Alkanes with five or more C-atoms have names based on Greek or Latin words for numbers, e.g. Greek: *pente* means five.

Checkpoint 1

Correct these names:
(a)   3, methyl-butane
(b)   4,5 dimethylheptane

Watch out!

Cyclohexane has the same molecular formula as hexene (C_6H_{12}) but it has no double bond because the six C-atoms form a ring.
As well as the molecular formula, organic compounds may be represented by a full displayed structural formula, an abbreviated structural formula or a skeletal formula.

The jargon

Isomers are different compounds with the same molecular formula.
cis- Greek same
trans- Greek across
E- German *entgegen* (opposite)
Z- German *zusammen* (together)
CIP rules named after Cahn, Ingold and Prelog who developed the E-Z system of nomenclature.

Checkpoint 2

Correct this name: 4, hexene. How many *straight-chain* hexenes, molecular formula C_6H_{12}, are there? What are their systematic names?

2. name the isomer *Z* if the higher priority groups are on the same side of the double bond.

3. name the isomer *E* if they are on opposite sides of the double bond.

 E-2-bromo-but-2-ene   *Z*-2-bromo-but-2-ene

Watch out!

cis-trans (and *E–Z*) isomers may have different physical and/or chemical properties, e.g.: Z butenedioic acid (b.p. 130°C) forms an anhydride on heating but E butenedioic acid (sublimes at 200°C) does not.

Naming arenes

Arenes are hydrocarbons containing benzene rings.

→ The simplest arene (C_6H_6) is called benzene.
→ The group (C_6H_5-) derived from it is called phenyl.
→ The carbon atoms are numbered to show the position of any groups attached to the ring.

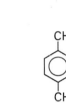

Watch out!

In benzene itself all six positions in the ring are the same.

Examples

methylbenzene (toluene)   1,2-dimethylbenzene (ortho-xylene)   1,3-dimethylbenzene (meta-xylene)   1,4-dimethylbenzene (para-xylene)

Checkpoint 3

Correct this name: 2,4,6-trimethylbenzene.
How many trichlorobenzenes are there? What are their systematic names?

Naming compounds with functional groups

The same rules apply to compounds containing functional groups such as halogeno (−Cl, −Br, −I), hydroxyl (−OH) and amino (−NH₂).

→ Name the functional groups in alphabetical order.

The jargon

A *functional group* is an element or combination of elements responsible for specific properties of an organic compound or class of compounds.

Example

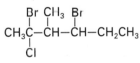

2,4-dibromo-2-chloro-3-methylhexane

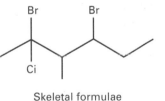

Skeletal formulae

Action point

Make yourself a set of cards to learn the names, structures and reactions of the important functional groups listed on pages 146–7.

Exam practice   answers: page 187

(a) Explain what is meant by the statement that 2-methylbutane and 2,2-dimethylpropane are *isomeric alkanes* and draw the structures of the two hydrocarbons.

(b) Give the systematic name for *each* of the following:

(i) CH₃   CH₃
     | |
     CH₃CHCH₂CHCH₃

(ii) CH₃CH=CHCH₂CH₃

(iii)

(c) Draw structures for each of the following:

(i) octane, (ii) cyclopentane, (iii) 3-chloroethylbenzene.

(d) Draw the three possible structures for the molecular formula $C_2H_2Cl_2$ and give the systematic name of each structure.

Examiner's secrets

If you are asked for a systematic name do *not* give a trivial name; you will lose marks. If you are just asked for the name of, say, a compound of formula CH_3CO_2H, *ethanoic acid* (systematic) or *acetic acid* (trivial) will do.

Classes of compounds and functional groups

To make sense of organic chemistry, we classify compounds according to the structure and properties of their molecules and functional groups. Here are the important classes, homologous series and functional groups for A-level.

Classes of organic compounds

We put compounds into broad classes (e.g. *aliphatic* and *aromatic*) which we divide into narrower classes (e.g. *hydrocarbons, oxygen-containing, nitrogen-containing*, etc.).

→ A homologous series is the simplest class of organic compound.
→ Alkanes are the simplest and most important homologues.

Alkanes (general formula C_nH_{2n+2})

CH_4	C_2H_6	C_3H_8	C_4H_{10}	$C_5H_{12} \ldots$
methane	ethane	propane	butane	pentane

→ Homologues have the same functional group(s).

Aliphatic primary alcohols (general formula $C_nH_{2n+1}OH$)

CH_3OH	C_2H_5OH	C_3H_7OH	C_4H_9OH	$C_5H_{11}OH \ldots$
methanol	ethanol	propan-1-ol	butan-1-ol	pentan-1-ol

Functional groups in hydrocarbons

Functional group		Prefix	Suffix	Class of compounds
>C:C<	>C=C<		ene	alkenes
C_6H_5-		phenyl	benzene	arenes and aromatics

Functional groups containing oxygen

Functional group		Prefix	Suffix	Class of compounds
—CHO	$-C\overset{H}{_{O}}$		al	aldehydes
>CO	>C=O	oxo	one	ketones and aldehydes
$-CO_2H$	$-C\overset{O}{_{O-H}}$		oic acid	carboxylic acids
—COCl	$-C\overset{O}{_{Cl}}$		oyl chloride	acid or acyl chlorides
$-CO_2$	$-C\overset{O}{_{O-}}$		oate	esters and polyesters
$(-CO)_2O$	$-C\overset{O}{_{O}}$ $-C\overset{O}{_{O}}$		anhydride	acid anhydrides

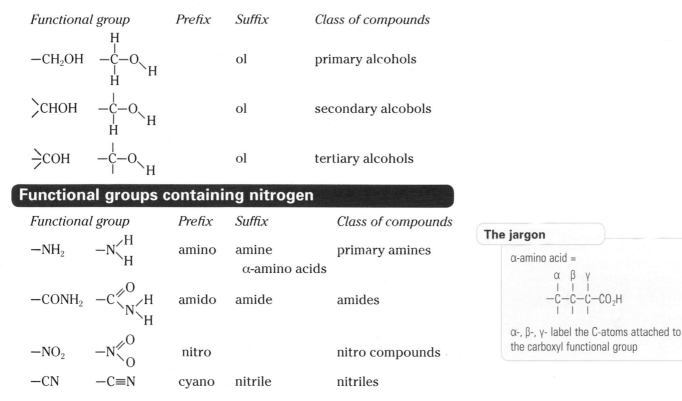

Functional group	Prefix	Suffix	Class of compounds
$-CH_2OH$ $-\overset{H}{\underset{H}{C}}-O_{\diagdown H}$		ol	primary alcohols
$>CHOH$ $-\overset{}{\underset{H}{C}}-O_{\diagdown H}$		ol	secondary alcohols
$>COH$ $-\overset{}{\underset{}{C}}-O_{\diagdown H}$		ol	tertiary alcohols

Functional groups containing nitrogen

Functional group	Prefix	Suffix	Class of compounds
$-NH_2$ $-N\overset{H}{\diagdown H}$	amino	amine α-amino acids	primary amines
$-CONH_2$ $-C\overset{O}{\underset{N\diagdown H}{\diagup H}}$	amido	amide	amides
$-NO_2$ $-N\overset{O}{\diagdown O}$	nitro		nitro compounds
$-CN$ $-C\equiv N$	cyano	nitrile	nitriles

➔ Proteins contain $-CO-NH-$ and are built from α-amino acids.

Functional groups containing halogen

Functional group	Prefix	Suffix	Class of compounds
-hal -hal	halogeno		halogeno compounds
$-COCl$ $-C\overset{O}{\diagdown Cl}$		oyl chloride	acid or acyl chlorides

The jargon

α-amino acid =

$$\underset{|}{\overset{\alpha}{C}}-\underset{|}{\overset{\beta}{C}}-\underset{|}{\overset{\gamma}{C}}-CO_2H$$

α-, β-, γ- label the C-atoms attached to the carboxyl functional group

The jargon

$-CO-NH-$ is an amide linkage

The jargon

hal = F, Cl, Br, I

Exam practice (8 minutes)   answers: page 107

(a) Write down *all* the functional groups you can identify in the compound known as CS, the structure of which is shown below.

(b) Draw the structures of
   (i) ethanoic acid, (ii) ethanol, (iii) ethyl ethanoate.

(c) Construct a balanced equation for the reaction of ethanol and ethanoic acid.

Grade booster

Here is a typical question you have been warned about and should be prepared for if you want the top grade. You may have heard about CS gas being used by the police but you aren't expected to have come across its formula. You can answer this question by applying your knowledge and understanding of structural formulae and functional groups. Use the tables on these two pages to answer part (a).

Types of reactions and reagents

There are four types of organic reaction: addition, elimination, rearrangement and substitution. They may involve polar bonds and electrons moving around. To understand the mechanism of a reaction, you must know about electrophiles, nucleophiles and free radicals; so what are they?

Types of reagent

Free radicals

When a single covalent bond (R—R) between two atoms with similar electronegativities breaks *homolytically*, each atom keeps one electron. The resulting reactive molecular fragments (R·) are called free radicals, e.g. ·Cl and ·CH_3.

→ A free radical is an atom (or group of atoms) with an unpaired electron.

Electrophiles and nucleophiles

When a single covalent bond (A—B) between two atoms with different electronegativities breaks *heterolytically*, one atom (A) loses both electrons and the other atom (:B) keeps both electrons. Fragment A is an electrophile, e.g. nitronium ion NO_2^+. Fragment :B is a nucleophile, e.g. hydroxide ion :OH^-.

→ An electrophile is an electron-pair acceptor (and a Lewis acid).
→ A nucleophile is an electron-pair donor (and a Lewis base).

Types of reaction

Free radical substitution

The reaction between an alkane and bromine in UV or strong sunlight is a chain reaction:

→ initiation   Br—Br → 2 ·Br
→ propagation   ·Br + H—CH_3 → H—Br + ·CH_3
                 Br—Br + ·CH_3 → Br—CH_3 + ·Br

The net effect of these two propagation steps is given by

$$Br—Br + H—CH_3 → H—Br + Br—CH_3$$

→ termination   ·CH_3 + ·CH_3 → CH_3—CH_3

CH_3Br can take part in the propagation steps to form CH_2Br_2. Substitution can continue until all four H-atoms in the methane molecule have been substituted.

Electrophilic addition

Propene reacts with hydrogen chloride to form 2-chloropropane:

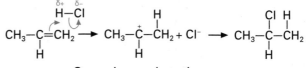

Secondary carbocation

The high electron density of the C=C bond makes alkenes react as nucleophiles and induces a polarity in approaching halogen molecules.

→ Addition occurs across the double bond to the two carbon atoms originally joined by it.

→ The addition product is 2-chloropropane, *not* 1-chloropropane, because the reaction follows Markovnikov's rule: when a molecule HZ adds to an unsymmetrical alkene, the H-atom adds to the double-bonded C-atom with the greater number of H-atoms.

Nucleophilic substitution

1-bromobutane is hydrolyzed slowly in *one step* by aqueous sodium

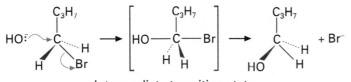

Intermediate transition state

hydroxide to form butan-1-ol.

As a lone pair of electrons on the hydroxide ion attacks the region of low electron density and begins to form a bond with the electrophilic carbon atom attached to the halogen, the carbon–halogen bond begins to break. In the middle of this process an unstable transition state is reached.

→ Primary halogenoalkanes favour the S_N2 mechanism

2-bromo-2-methylpropane is hydrolyzed rapidly in *two steps* by aqueous sodium hydroxide to form 2-methylpropan-2-ol.

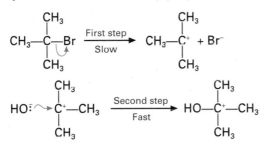

→ Tertiary halogenoalkanes favour the S_N1 mechanism

> **Exam practice   10 minutes**   answers: page 187

(a) For the photochemical chlorination of ethane, write balanced equations to show (i) the initiation reaction, (ii) *two* propagation reactions, (iii) a termination reaction, and (iv) the *overall* reaction showing $C_2H_4Cl_2$ as the organic product.

(b) Suggest why the frequency of light needed for free radical halogenation of alkanes is higher for chlorine than for bromine.

(c) Suggest how a CFC such as CCl_2F_2 might destroy ozone and explain why depletion of the ozone layer is considered harmful.

The jargon

A *carbocation* (or *carbonium ion*) is an organic cation in which a carbon atom carries a positive charge. *Markovnikov's* rule can be explained in terms of the stabilities of carbocations.

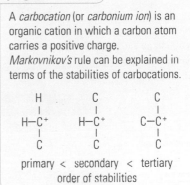

primary < secondary < tertiary
order of stabilities

Checkpoint 3

(a) What type of reagent is (i) a carbocation, (ii) a chloride ion? (b) Name the product and explain the mechanism for the reaction of bromine vapour with ethene gas.

The jargon

S_N2 means a bimolecular (2) nucleophilic (N) substitution (S) reaction. The '2' means the rate-determining step involves two species (the nucleophile and the electrophile). Rate equation is rate = $k[\text{RBr}][\text{OH}^-]$.

The jargon

S_N1 means a unimolecular (1) nucleophilic (N) substitution (S) reaction. The '1' means the rate-determining step involves one species (the electrophile). Rate equation is rate = $k[\text{RBr}]$.

Watch out!

The CH_3 group in methylbenzene reacts like methane with free radicals.

Hydrocarbons: alkanes and alkenes

Hydrocarbons are compounds of carbon and hydrogen only. They may be saturated or unsaturated and their molecules may contain chains and/or rings of carbon atoms.

The jargon

Saturated means there are only single C—C bonds. *Unsaturated* means there are multiple bonds (e.g. double C=C) in chains and rings.

Checkpoint

Predict and draw the shape of an ethane molecule.

Grade booster

To see if you understand basic ideas you may be asked to suggest why the b.p. of $C(CH_3)_4$ (= 9.5°C) is lower than b.p. of $CH_2CH_2CH_2CH_2CH_3$ (= 36°C). Revise van der Waals' forces on page 18 if you want the top grade.

The jargon

A *radical* or *free radical* is a species with an unpaired electron. These radicals are very reactive.

The jargon

Thermal cracking is by heat. *Catalytic cracking* is by heat and heterogeneous catalysts.

The jargon

−1−, −2−, −3− is the position of double bond.

Alkanes

→ Simplest homologous series ($C_nH_{(2n+2)}$)

The first four are petroleum gases and methane is also natural gas. Alkanes are also constituents of diesel, petrol, aviation fuel, oils, etc.

→ Nomenclature of the other homologous series is based on alkanes

The melting and boiling points of the alkanes increase as molar mass increases (because van der Waals forces increase as *n* increases). Boiling point decreases as the alkane chain becomes more branched.

Reactions of alkanes

The alkanes undergo very few reactions.

Combustion

Alkanes can burn to water and carbon dioxide (or carbon monoxide if insufficient air): $CH_4 + 2O_2 \rightarrow CO_2 + 2H_2O$.

Chlorination

→ Alkanes undergo substitution reactions with chlorine in the presence of UV light.
→ UV light causes homolytic fission of chlorine molecules and the formation of free radicals.

1 $Cl_2 + h\upsilon \rightarrow 2Cl\cdot$ chain initiation
2 $Cl\cdot + CH_4 \rightarrow \cdot CH_3 + HCl$ ⎫
3 $\cdot CH_3 + Cl_2 \rightarrow CH_3Cl + Cl\cdot$ ⎬ chain propagation
4 $\cdot CH_3 + \cdot CH_3 \rightarrow C_2H_6$ chain termination
5 Further substitution can give CH_2Cl_2, $CHCl_3$ and CCl_4.

Cracking

Smaller alkane molecules are more useful than some of the larger ones obtained from the primary distillation of petroleum.
→ Alkanes are split by thermal and catalytic cracking into smaller alkane molecules and some unsaturated molecules:

$$C_{14}H_{30} \rightarrow C_{10}H_{22} + 2C_2H_4$$

→ Unsaturated hydrocarbons are used to make synthetic polymers.

Alkenes

→ Unsaturated hydrocarbons (C_nH_{2n}) starting at $n = 2$
→ Names are based on the alkanes ('a' changes to 'e')

$CH_2{=}CH_2$ $CH_3CH{=}CH_2$ $CH_3CH{=}CHCH_3$ $C_3H_7CH{=}CH_2$
ethene   propene   but-2-ene(s)   pent-1-ene
$C_4H_9CH{=}CH_2$ $C_3H_7CH{=}CHCH_3$ $C_2H_5CH{=}CHC_2H_5$
hex-1-ene   hex-2-ene   hex-3-ene

Reactions of alkenes

Alkenes combust to CO_2 and H_2O but they also react by *addition*.

Halogen addition

→ Alkenes rapidly 'decolourize' bromine (brown) in organic solvents: $CH_2=CH_2 + Br_2 \rightarrow CH_2Br{-}CH_2Br$
   or in water: $CH_2=CH_2 + Br_2 + H_2O \rightarrow CH_2Br{-}CH_2OH + HBr$

The $C=C$ double bond is also rapidly oxidized by warm alkaline aqueous potassium manganate(VII). In this test the purple colour disappears and a brown precipitate is formed in the colourless solution.

Catalytic hydrogenation

→ Hydrogen reduces alkenes to alkanes by adding across the double bond to form a single bond: $CH_2=CH_2 + H_2 \rightarrow CH_3{-}CH_3$.
→ Transition metals like platinum, palladium and nickel (the cheapest) are used commercially as catalysts.

Mechanism of addition reactions

→ Reaction of bromine (or hydrogen bromide) with an alkene is an *electrophilic addition*.

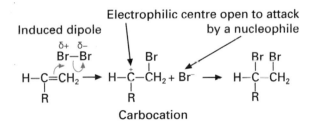

If bromine is in water or NaCl(aq), other nucleophiles can compete with the bromide ion and attack the electrophilic carbon atom. For example, a water molecule (instead of a bromine molecule) may attack the $C=C$ bond and release the nucleophilic OH^- ion.

→ When HBr adds to an unsymmetrical alkene, Br adds to the carbon with the smaller number of H-atoms (Markovnikov's rule).
→ Propene forms 2-bromopropane (*not* 1-bromopropane) in an electrophilic addition reaction with HBr.

Exam practice   (10 minutes)   answers: page 188

(a) Explain the difference between electrophilic addition and nucleophilic substitution.

(b) Show the mechanism for the chlorination of ethene to form 1,2-dichloroethane and state the conditions for the reaction.

(c) Give the names and structural formulae of *three* possible organic compounds formed when ethene is passed into an aqueous solution of bromine and sodium chloride.

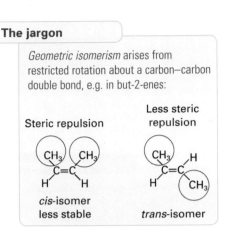

The jargon

Geometric isomerism arises from restricted rotation about a carbon–carbon double bond, e.g. in but-2-enes:

Steric repulsion   Less steric repulsion

cis-isomer   *trans*-isomer
less stable

Grade booster

If you are asked to give a laboratory test for cyclohexene, you should describe how to do the Br2 or KMnO4 test *and* what to observe for a positive result. Cyclohexene is a liquid but does the same thing as ethene.

Don't forget!

Addition reactions turn a double bond into a single bond. If you leave the double bond unaltered and write a formula like $H_3C=CH_2Br$ with a carbon atom having five bonds, you will lose marks and the top grade.

The jargon

Markovnikov's rule is
$HX + R_2HC=CH_2 \rightarrow R_2HCX{-}CH_2R$
because the order of carbonium ion stability is tertiary > secondary > primary or $R_3C^+ > R_2HC^+ > RH_2C^+$.

Hydrocarbons: alkenes and arenes

Saturated hydrocarbons are dull compared to unsaturated compounds like the ethene and benzene. The double bond lets us turn alkenes into alcohols, polymers, etc., by addition reactions. The delocalized benzene ring lets us turn arenes into azo dyes, explosives, etc., by substitution reactions. So how do we make polymers from alkenes?

The jargon

Greek: *polys* – many.
Poly(ethene), poly(chloroethene), etc. are named using the prefix 'poly' + (monomer name).
In *addition polymerization* monomers combine without the elimination of any other molecules.
In *condensation polymerization* monomers combine with the elimination of other (small) molecules.

Grade booster

You will definitely lose marks and miss the top grade if you make the common mistake of representing the repeating structure of poly(ethene) like

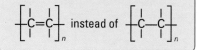

Checkpoint 1

Draw a diagram to show the formation of poly(propene).

Watch out!

Poly(2-methylpropenoate) or polymethylmethacrylate = 'perspex' is also an addition polymer (*not* a condensation polymer or polyester).

Don't forget!

The catalytic hydrogenation of alkenes (and unsaturated oils) to alkanes (and saturated fats).

The jargon

Ethane-1,2-diol = ethylene glycol = glycol = antifreeze.

Manufacture of synthetic polymers from alkenes

→ Ethene polymerizes to poly(ethene) or 'polythene'.
→ Polymerization occurs when a large number of small molecules (monomers) combine to form a large molecule (polymer).
→ Alkenes (and their derivatives) are used to form synthetic polymers.
→ **Low-density poly(ethene)** is formed by heating ethene under pressure with a *trace* of oxygen: $nCH_2=CH_2 \rightarrow (-CH_2-CH_2)_n-$.
→ **High-density poly(ethene)** is made at much lower pressures and temperatures using Ziegler–Natta catalysts (e.g. $Al(C_2H_5)_3$ and $TiCl_4$).

The high-density polymer is tougher and denser than the low-density product because it is more crystalline.

There is a whole range of polymers based on a substituted ethene monomer. The repeat unit can be written as $-CH_2-CHX-$ (where X could be for example H, Cl, CH_3, C_6H_5, etc.) and the polymerization could be represented as follows:

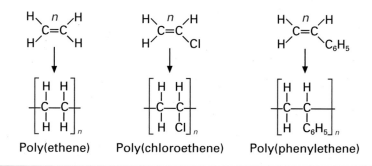

Poly(ethene)   Poly(chloroethene)   Poly(phenylethene)

Manufacture of alcohols from alkenes

Ethene is an important industrial feedstock in the large-scale manufacture of a variety of chemicals:

1 direct hydration to ethanol

$$C_2H_4(g) + H_2O(g) \xrightarrow[H_3PO_4 \text{ catalyst}]{70 \text{ atm } 300°C} C_2H_5OH(g)$$

2 indirect hydration to ethane-1,2-diol

$$C_2H_4(g) + {}^{1}/_{2}O_2(g) \xrightarrow[\text{silver catalyst}]{} \underset{\text{epoxyethane}}{CH_2-CH_2(g)}$$

$$\underset{O}{CH_2-CH_2}(g) + H_2O(g) \xrightarrow{2 \text{ atm } 150°C} CH_2OH-CH_2OH(g)$$

Ethane-1,2-diol is one of the monomers in the production of the polyester 'Terylene'.

Benzene

Benzene does *not* react like an alkene with three C=C bonds and is $150 \, \text{kJ mol}^{-1}$ more stable than Kekulé's structure predicts. In fact the 12 atoms are coplanar: the six C-atoms form a regular hexagon and the C–C bond length is between C—C and C=C.

 Kekulé structure consistent with hydrogenation of C_6H_6 to C_6H_{12}

 Molecular orbital model with sideways overlap of p-orbitals forming the delocalized π-electrons above and below the ring

 now represents the delocalized structure of benzene

Watch out!

If you want the top grade don't forget the H-atoms attached to the ring.

Benzene is C_6H_6

C_6H_5 is the phenyl group

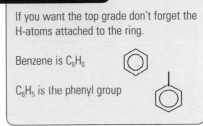

Reactions of arenes

The π-electrons above and below the ring make benzene open to attack by electrophiles.

→ Arenes (nucleophiles) mainly undergo substitution reactions.

→ These reactions are *electrophilic substitution reactions*.

Nitration of benzene

Benzene reacts with a nitrating mixture of conc. nitric acid in conc. sulphuric acid to form nitrobenzene.

1　Nitronium ions are formed in the nitrating mixture:

$$2H_2SO_4 + HNO_3 \rightarrow NO_2^+ + H_3O^+ + 2HSO_4^-$$

2　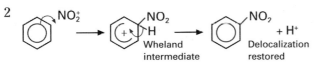

Wheland intermediate　　Delocalization restored

3　$H^+ + HSO_4^- \rightarrow H_2SO_4$

Overall reaction: $C_6H_6 + HNO_3 \rightarrow C_6H_5NO_2 + H_2O$

The jargon

Nitronium ion = nitryl cation = NO_2^+.
Wheland intermediate – cation with four π-electrons shared by five carbon atoms in the benzene ring.

Chlorination of methylbenzene

→ Chlorine reacts with the side chain in UV light and reacts with the benzene ring in the absence of light (and heat) but in the presence of a carrier (e.g. $AlCl_3$).

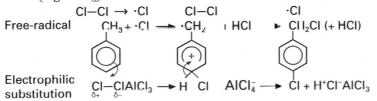

Free-radical　$Cl—Cl \rightarrow \cdot Cl$
$CH_3 + \cdot Cl \longrightarrow \cdot CH_2 + HCl$　$\cdot Cl \blacktriangleright CH_2Cl \, (+ HCl)$

Electrophilic substitution　$Cl—Cl AlCl_3 \rightarrow H \; Cl$　$AlCl_4^- \longrightarrow Cl + H^+Cl^-AlCl_3$

→ Free radical chlorination of the methyl group can replace all three H-atoms to give $C_6H_5CH_2Cl$, $C_6H_5CHCl_2$ and $C_6H_5CCl_3$.

Nitration of the ring can give 2,4,6-trinitrobenzene (TNT).

The jargon

A *side chain* is a methyl group or any other alkyl group attached to the ring.

The jargon

methylbenzene = toluene
TNT = trinitro toluene

Checkpoint 2

Write the structure of TNT and a balanced equation for its formation from methylbenzene.

Exam practice　(5 minutes)　answers: page 188

(a) Write an equation for the chlorination of methylbenzene to
　(i) (trichloromethyl)benzene and (ii) 1-chloro-2-methylbenzene.

(b) Describe and explain the mechanism of the reaction between methylbenzene and chlorine in the presence of UV light.

Compounds containing halogens

In organic compounds, halogens are joined to carbon by single covalent bonds. The strength of the bond depends on the halogen and on the other atoms joined to the carbon. C—F is strongest and C—I weakest. C_6H_5Cl is unreactive and CH_3COCl very reactive.

The jargon

A *halogenoalkane* has a halogen substituted into an alkane chain.

The jargon

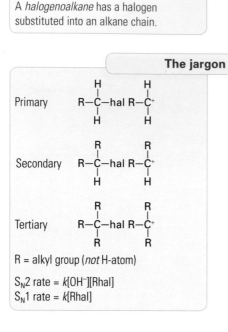

R = alkyl group (*not* H-atom)

S_N2 rate = $k[OH^-][Rhal]$
S_N1 rate = $k[Rhal]$

Links

See page 43: hydrolysis of bromoalkanes.

Grade booster

You don't need to remember the E_N2 mechanism but you should know how aqueous and alcoholic alkali differ in their reaction on halogenoalkanes if you want the top grade.

The jargon

Alcoholic means in aqueous ethanol. A preferred word would be *ethanolic*.

Watch out!

When a C=C bond forms, the H-atom eliminated comes from the C-atom next to the carbon bonded to the halogen. However, sometimes a cycloalkane (no C=C bond) is formed.

Halogenoalkanes

→ Iodo compounds tend to be the most reactive because the C—I bond is the weakest (relative bond strength is C—Cl> C—Br> C—I). Halogenoalkanes are attacked by nucleophiles, undergoing substitution or elimination reactions depending upon the reagent and conditions.

→ The electrophilic centre is the carbon atom attached to the halogen.

Alkaline hydrolysis

→ Nucleophilic *substitution* with *aqueous* sodium hydroxide
→ S_N2 reaction mechanism for *primary* halogenoalkanes

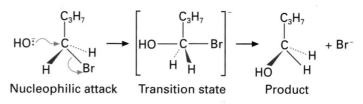

Nucleophilic attack   Transition state   Product

A similar S_N2 mechanism applies to the nucleophilic substitution of *primary* halogenoalkanes by CN^- (or NH_3) to form nitriles (or amines). Kinetics experiments indicate that tertiary (and some secondary) halogenoalkanes react by a different mechanism.

→ S_N1 reaction mechanism for *tertiary* halogenoalkanes

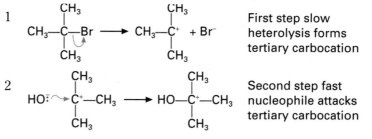

1. First step slow heterolysis forms tertiary carbocation

2. Second step fast nucleophile attacks tertiary carbocation

→ Nucleophilic *elimination* with *alcoholic* potassium hydroxide
→ E_N2 reaction mechanism for *primary* halogenoalkanes

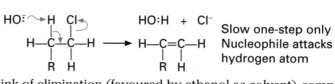

Slow one-step only Nucleophile attacks hydrogen atom

You can think of elimination (favoured by ethanol as solvent) competing with substitution (favoured by water as solvent).

→ Elimination reactions occur with tertiary halogenoalkanes more readily than with primary halogenoalkanes.

Aromatic halogen compounds

→ Halogens attached to the benzene ring are not reactive because their p-electrons become part of the ring's delocalized π-system.
→ Halogen attached to the side chain behaves like halogenoalkane.

Acyl chlorides

The two compounds you meet at A-level are ethanoyl chloride (CH_3COCl) and benzoyl chloride (C_6H_5COCl).

→ Acyl chlorides are very reactive.

Electronegative oxygen and chlorine make the attached carbon strongly electrophilic and readily attacked by nucleophiles

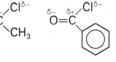

→ Acyl chlorides fume in moist air: $RCOCl + H_2O \rightarrow RCO_2H + HCl$.
→ $CH_3COCl(l)$ and $C_6H_5COCl(l)$ are powerful acylating agents used, for example, to make esters and amides.

Phenol (unlike alcohols) will not react directly with carboxylic acids to form esters.

→ Phenol will react with acyl chlorides to form esters.

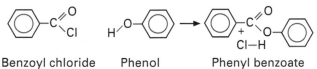

Benzoyl chloride   Phenol   Phenyl benzoate

Uses of halogen compounds

Their commercial and industrial importance depends upon their ability to take part in a variety of nucleophilic substitution reactions.

→ As intermediates in synthetic organic chemistry: e.g. extending the carbon chain $ROH \rightarrow RBr \rightarrow RCN \rightarrow RCO_2H \rightarrow RCH_2OH$.
→ Chloroethane was used to make tetraethyl lead (main antiknock agent in leaded petrol).
→ 1,2-dibromoethane was used as a lead scavenger in the fuel.
→ Chlorofluorocarbons have been used as refrigerants and as propellants in aerosols.
→ Chloroethene is the monomer for making PVC.
→ Chlorinated hydrocarbons such as tetrachloromethane, 1,1,1-trichlorethane and trichloromethane are good solvents.
→ Chlorobenzene is used in the manufacture of DDT.

Environmental concerns

The EU has banned CFCs to avert further damage to the ozone layer. The use of the insecticide DDT against malaria-carrying mosquitoes is restricted because it dissolves in animal fat and accumulates in our food chain. Vinyl chloride is very toxic and disposal of PVC by burning produces highly toxic chlorinated dioxins. Chlorinated solvents can irreversibly damage internal organs if they are ingested or inhaled. Some, like CCl_4, are dangerously carcinogenic. Some volatile solvents are being replaced by ionic liquids.

The jargon

Acyl group = R—CO— =$R \atop \rangle C=O$
R = alkyl or aryl
R ≠ H, OH, NH_2
In acylation RCO is substituted for H in —OH, —NH_2, etc.
Substituting CH_3CO— is ethanoylation.

Links

See page 159: formation of esters by carboxylic acids.

Checkpoint

Draw structures of the products for ethanoyl chloride reacting with
(i) ethanol
(ii) phenol
(iii) phenylamine
(iv) ethylamine

The jargon

Chloroethene or vinyl chloride = CH_2=CHCl
Trichloromethane or chloroform = $CHCl_3$
1,1,1-trichlorethane = CH_3CCl_3
1,1,1-trichloro-2,2-di(4-chlorophenyl)ethane = DDT =

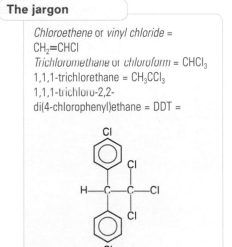

Exam practice (3 minutes)   answers: page 188

Draw and explain the mechanism for the substitution reaction of potassium cyanide with 1-bromopropane.

Compounds containing oxygen: alcohols, phenols,

For A-level you study alcohols and phenol, aldehydes and ketones, carboxylic acids and their derivatives.

The jargon

Functional groups:
Alcohol = OH
Phenol = C_6H_5OH
Aldehyde = RCHO
Ketone = R_2CO
Carboxylic acid = RCOOH or RCO_2H
Ester = RCOOR' or RCO_2R' R' ≠ H
Acyl chloride = RCOCl
Amide = $RCONH_2$

Checkpoint 1

Write the structures and identify each class (*p-*, *sec-* and *tert-*) of the alcohol of formula C_4H_9OH.

Alcohols

→ Hydroxyl (—OH) is the functional group of alcohols.
→ OH attached to a benzene ring is the functional group of phenols.

We class alcohols as *primary*, *secondary* or *tertiary* according to the number of H-atoms attached to the carbon atom of the C—OH group.

Aliphatic alcohols

CH_3OH	methanol	$C_2H_5CH_2OH$	propan-1-ol
C_2H_5OH	ethanol	$CH_3CH(OH)CH_3$	propan-2-ol
$C_3H_7CH_2OH$	butan-1-ol	$C_2H_5CH(OH)CH_3$	butan-2-ol

Formation of alcohols

→ In general alcohols are formed by hydrolysis of halogenoalkanes with the bromo and iodo compounds giving the best yields:

$C_2H_5I + NaOH(aqueous) \rightarrow C_2H_5OH + NaI$ [heat under reflux]

→ Ethanol is made industrially by the direct hydration of ethene

$C_2H_4 + H_2O \rightarrow C_2H_5OH$ (70 atm, 300 °C, phosphoric acid catalyst)

and by fermentation of carbohydrates

$C_6H_{12}O_6 \rightarrow 2C_2H_5OH + 2CO_2$ (catalyzed by enzymes from yeast)

Reactions of alcohols

→ Alcohols react with metallic sodium to give hydrogen gas:

$$C_2H_5OH + Na \rightarrow C_2H_5O^-Na^+ + {}^{1}\!/_{2}H_2$$

Watch out!

Oxidizing agents affect *sec-* and *tert-* alcohols differently, e.g. acidified dichromate oxidizes *sec-*alcohols to ketones but does not oxidize tertiary alcohols. Bear this in mind if you want the top grade.

→ Alcohols are oxidized by acidified potassium dichromate(VI)

$C_2H_5OH \rightarrow CH_3CHO$ (continuously distilled from oxidant)
$C_2H_5OH \rightarrow CH_3CHO \rightarrow CH_3COOH$ (refluxed with oxidant)

→ Alcohols form esters when refluxed with carboxylic acids and a mineral acid catalyst:

The jargon

Polyhydric alcohols have two or more OH groups in the molecule.
Ethan-1,2-diol = CH_2OHCH_2OH
Propan-1,2,3-triol = glycerol = $CH_2OHCHOHCH_2OH$

$$CH_3C{\overset{O}{\underset{OH}{\diagup\!\diagdown}}} + C_2H_5OH \xrightarrow{\text{ethano}} CH_3C{\overset{O}{\underset{OC_2H_5}{\diagup\!\diagdown}}} + H_2O$$

ethanoic acid   ethyl ethanoate   ethyl

→ Alcohols react with PCl_5 to give fumes of hydrogen chloride:

$C_2H_5OH + PCl_5 \rightarrow C_2H_5Cl + POCl_3 + HCl$ (test for OH groups)

Uses of alcohols

→ Methanol: manufacture of 'Perspex', methanoic and ethanoic acids
→ Ethanol: fuel, solvent and ingredient in wines, beers and spirits
→ Propan-2-ol: solvent and manufacture of propanone, phenol, etc.
→ Ethan-1,2-diol: antifreeze and manufacture of polyesters

Phenols

→ C_6H_5OH is the simplest phenol and a slightly water-soluble weak acid (carbolic acid) – conjugate base is $C_6H_5O^-$ (phenoxide ion):

$C_6H_5OH + H_2O \rightleftharpoons C_6H_5O^- + H_3O^+$; $K_a = 1.3 \times 10^{-10}$ mol dm^{-3}

aldehydes and ketones

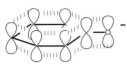

phenoxide ion stabilized by sideways overlapping p-orbitals of oxygen becoming part of the ring's delocalized π-system

→ Phenol dissolves in aqueous sodium hydroxide to form a solution of sodium phenoxide: $C_6H_5OH + NaOH \rightarrow C_6H_5ONa + H_2O$.
→ Phenol is weaker than carbonic acid, so carbon dioxide displaces it from the salt: $C_6H_5ONa + CO_2 + H_2O \rightarrow C_6H_5OH + NaHCO_3$.

Differences between phenol and ethanol

→ Aqueous phenol is acidic.
→ Phenol does not react with hydrogen halides, PBr_3 or PI_3.
→ Phenol resists oxidation but strong oxidants produce a complex mixture of aliphatic and aromatic products.

Tests for phenol

→ Bromine water gives a white precipitate (2,4,6-tribromophenol).
→ Aqueous iron(III) chloride gives characteristic violet colours with most phenolic compounds.

Uses of phenols

→ TCP (trichlorophenol) is a mixture of chlorinated phenols
→ manufacture of phenolic (Bakelite) and epoxy resins
→ manufacture of cyclohexanol for producing nylon

Aldehydes and ketones

For A-level, ethanal, CH_3CHO, and propanone, $(CH_3)_2CO$, are the most important carbonyl compounds. Methanal (formaldehyde), HCHO, is the simplest aldehyde but it has some atypical properties.

Aldehydes

→ Laboratory preparation by oxidation of alcohols controlled to avoid further oxidation to carboxylic acids.
→ Catalytic oxidation of ethanol vapour over copper catalyst.
→ Commercial manufacture of ethanal from
  1   ethene and steam using palladium(II) chloride catalyst
  2   ethanol by air oxidation and dehydrogenation with silver catalyst: $CH_3CH_2OH + \frac{1}{2}O_2 \rightarrow CH_3CHO + H_2O$ (exothermic) gives heat to $CH_3CH_2OH \rightarrow CH_3CHO + H_2$ (endothermic).

The jargon

These are compounds in which the —OH group is attached directly to the benzene ring. The simplest phenol is phenol itself, C_6H_5OH.

Checkpoint 2

Write an ionic equation for the reaction of phenol with NaOH(aq).

The jargon

2,4,6-tribromophenol =

Watch out!

Iron(III) chloride test works best in 'neutral' solution: too acidic – too little $C_6H_5O^-$ ion; too alkaline – too little Fe^{3+}.

The jargon

A *carbonyl group* is $>C=O$. *Carbonyl compounds* are

$$\begin{array}{c} R \\ H \end{array} \!\! \overset{\delta+}{\underset{}{}} C \overset{\delta-}{=} O \qquad \begin{array}{c} R \\ R \end{array} \!\! \overset{\delta+}{\underset{}{}} C \overset{\delta-}{=} O$$

aldehydes   and   ketones (R ≠ H)

Watch out!

This catalytic oxidation of $C_2H_5OH \rightarrow CH_3CHO$ needs a fume cupboard.

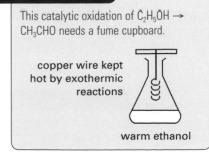

copper wire kept hot by exothermic reactions

warm ethanol

Exam practice   (10 minutes)   answers: page 189

(a) State how and under what conditions ethanol reacts with acidified dichromate.

(b) Explain how (i) ethanol is manufactured from ethene by direct hydration, (ii) glycerol is formed by the alkaline hydrolysis of oils and fats.

Compounds containing oxygen: aldehydes, ketones

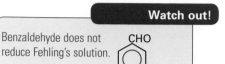
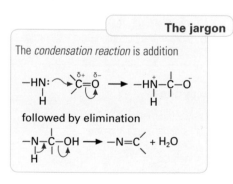
The key to the chemistry of this group is the polarity of the >C=O group.

Reactions of aldehydes

→ Aldehydes (but *not* ketones) are reducing agents identified by the following popular tests:

Test	*Positive observation*
→ Warm with Fehling's solution: complex Cu(II) ions in alkali	→ Colour changes from blue to green and finally get red precipitate Cu_2O
→ Warm with Tollens' reagent: complex Ag(I) ions in alkali	→ Silver mirror forms on inside of test tube or black precipitate deposited
→ Warm with acidified aqueous potassium dichromate	→ Colour of solution turns from orange to green

Reactions of ketones

→ Ketones are *not* readily oxidized.

Methyl ketones (and compounds easily oxidized to a methyl ketone) will produce a yellow precipitate of triiodomethane (iodoform) when warmed with aqueous sodium hydroxide and iodine (or aqueous sodium chlorate(I) and aqueous potassium iodide). The *iodoform reaction* is a test for CH_3CO- group or a group, like $CH_3CH(OH)-$, easily oxidized to it.

Reactions of aldehydes and ketones

→ The carbonyl group >C=O is attacked by nucleophiles.

Condensation reactions

→ Carbonyl compounds react with 2,4-dinitrophenylhydrazine to form yellow/orange precipitates of 2,4-dinitrophenylhydrazones:

Addition reactions

→ HCN and HSO_3^- add to >C=O to give $>C(OH)CN$ and $>C(OH)SO_3^-$:

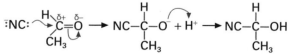

aqueous sodium or potassium cyanide is followed by excess mineral acid

Reduction

→ Sodium tetrahydridoborate(III), $NaBH_4$, reduces aldehydes to *primary* alcohols and ketones to *secondary* alcohols:
ethanal CH_3CHO [+ 2H] → CH_3CH_2OH ethanol
propanone $(CH_3)_2CO$ [+ 2H] → $(CH_3)_2CHOH$ propan-2-ol

and carboxylic acids

Carboxylic acids

Methanoic acid and ethanoic acid are the first two homologues of the straight-chain aliphatic monobasic weak carboxylic acids.

$$CH_3-C\overset{O}{\underset{O-H}{}} + :\overset{H}{\underset{H}{O}} \rightleftharpoons \left[CH_3-C\overset{O}{\underset{O}{}}\right]^- + \left[H-\overset{H}{\underset{H}{O}}\right]^+$$

Ethanoate ion stabilized by delocalization

→ Ethanoic acid is stronger than carbonic acid and (unlike phenol) liberates CO_2 from carbonates and hydrogencarbonates:
$CH_3CO_2H + NaHCO_3 \rightarrow CH_3CO_2Na + H_2CO_3 \rightarrow H_2O + CO_2$

Aromatic carboxylic acids like benzoic acid and 2-hydroxybenzoic acid dissolve in hot water but less well than aliphatic acids in cold water.

Reactions of carboxylic acids

→ Lithium tetrahydridoaluminate(III), LiAlH$_4$, in dry ether reduces the carboxyl group to primary alcohol: RCOOH → RCH$_2$OH
→ PCl$_5$ (or SOCl$_2$) attack the OH in the carboxyl group to form acyl chloride: RCOOH + PCl$_5$ → RCOCl + POCl$_3$ + HCl
→ Refluxing with alcohols and a mineral acid catalyst (H$_2$SO$_4$ or HCl) esterifies the carboxyl group to form sweet smelling neutral compounds: CH$_3$COOH + C$_2$H$_5$OH ⇌ CH$_3$COOC$_2$H$_5$ + H$_2$O
ethyl ethanoate (an ester)
→ Decarboxylation (heating with soda lime: CaO(s) + NaOH(aq)) forms hydrocarbons: RCO$_2$H → RH + CO$_2$ (→ CaCO$_3$; Na$_2$CO$_3$)

Carboxylic acid derivatives

Acyl chlorides and acid anhydrides

→ Reactive compounds readily hydrolyzed by water back to carboxylic acids: CH$_3$COCl + H$_2$O → CH$_3$COOH + HCl
→ React with (acylate) OH and NH$_2$ groups replacing an H with an acyl group (RCO): CH$_3$COCl + HOCH$_3$ → CH$_3$COOCH$_3$ + HCl

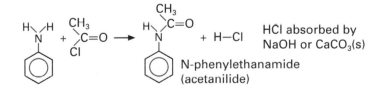

HCl absorbed by NaOH or CaCO$_3$(s)

N-phenylethanamide (acetanilide)

Amides and esters

→ On refluxing with NaOH(aq) amides give ammonia and esters give the alcohol and sodium salt of the carboxylic acid:
ethanamide: CH$_3$CONH$_2$ + NaOH → CH$_3$CO$_2$Na + NH$_3$
ethyl methanoate: HCO$_2$C$_2$H$_5$ + NaOH → HCO$_2$Na + C$_2$H$_5$OH

Exam practice   (6 minutes)   answers: page 109

(a) Write an equation for ethanoyl chloride reacting with 2-hydroxybenzoic acid. Draw the structure of the organic product. (b) How could you prepare methyl 2-hydroxbenzoate (oil of wintergreen) from 2-hydroxybenzoic acid.

The jargon

The *carboxyl group* is —COOH =

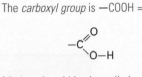

Methanoic acid is also called *formic acid*.
Ethanoic acid is also called *acetic acid*.
Esterification is the formation of esters from carboxylic acid and alcohol.

Checkpoint 2

Write an equation for the formation of methyl propanoate.

The jargon

The acetyl derivative of salicylic acid is called *aspirin*.

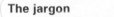

The jargon

Ethanoyl chloride = CH$_3$COCl.
Acylation with RCO = CH$_3$CO is ethanoylation.

The jargon

The 'N' in N-phenylethanamide means the phenyl group is attached to the nitrogen atom.

Compounds containing nitrogen: amines and

All A-level syllabuses refer to amines, amides, polyamides, amino acids and proteins. The key to your study of these and other related substances is the amino group, $-NH_2$, and the amide (peptide) link $-CO-NH-$.

Checkpoint 1

Write the structural formula of the following and identify each as a primary, secondary or tertiary amine:
(i)   2-aminobutane
(ii)   2-methylphenylamine
(iii) ethyldimethylamine
(iv)   N:N-dimethylphenylamine
Hint: N:N means both methyl groups attached to the nitrogen atom of the amine group.

Amines

Amines are derivatives of ammonia classified by the number of carbons attached to the nitrogen atom:

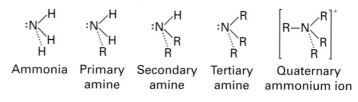

Ammonia — Primary amine — Secondary amine — Tertiary amine — Quaternary ammonium ion

→ Ammonia and amines are *weak* bases (proton acceptors) and nucleophiles because of the lone electron pair on the N-atom:
$NH_3 + H_2O \rightleftharpoons NH_4^+ + OH^-$ and $RNH_2 + H_2O \rightleftharpoons RNH_3^+ + OH^-$
→ Ethylamine is a slightly stronger base and phenylamine a slightly weaker base than aqueous ammonia.

Formation of primary amines

→ Nucleophilic substitution reactions of ammonia with halogenoalkanes: $NH_3 + Rhal \rightarrow RNH_2 + Hhal$ ($\rightarrow NH_4hal$).
→ Reduction of nitriles and amides with lithium tetrahydridoaluminate(III): $RCN\ [+4H] \rightarrow RCH_2NH_2$; $RCONH_2\ [+4H] \rightarrow RCH_2NH_2 + H_2O$.
→ Reduction of aromatic nitro compounds, e.g. nitrobenzene $C_6H_5NO_2[+6H] \rightarrow C_6H_5NH_2 + 2H_2O$ to phenylamine.

With nitrobenzene you could use tin and concentrated hydrochloric acid as the reducing agent but the excess acid would form a salt with phenylamine: $C_6H_5NH_2 + HCl \rightarrow C_6H_5NH_3^+ + Cl^-$. So you must add excess NaOH to free the amine and then extract it from the product mixture by *steam distillation*.

Reactions of primary amines

→ NH_2 group reacts with (and during a synthesis protected by) acyl chlorides: $RNH_2 + CH_3COCl \rightarrow RNHCOCH_3 + HCl$.

The primary NH_2 group reacts with nitrous acid (from HCl(aq) and sodium nitrite, $NaNO_2$) to form unstable diazonium ions:

$$NH_{3(aq)}^+ + HNO_2(aq) \rightarrow RN_2^+(aq) + 2H_2O(l)$$

With *aliphatic* amines, even in ice-cold conditions, nitrogen gas is liberated almost quantitatively, $RN_2^+(aq) \rightarrow R^+(aq) + N_2$, and the very unstable electrophilic carbocation combines with various nucleophiles giving a mixture of organic products including RCl, RNO_2, ROH, etc. If conditions are *not* ice-cold, a similar reaction occurs with *aromatic* primary amines: $C_6H_5NH_2 + HNO_2(aq) \rightarrow C_6H_5OH + H_2O + N_2$.

→ If conditions are ice-cold, the *diazotization* of an *aromatic primary amine* forms a *stable aqueous aromatic diazonium cation*.

Watch out!

When ammonia and halogenoalkanes are heated in sealed tubes the reaction produces a mixture of *p-*, *sec-* and *tert-* amines.

The jargon

tetrahydridoaluminate(III) = $LiAlH_4$ = lithium aluminium hydride.

Grade booster

There are often questions about the preparation of phenylamine from nitrobenzene because the process involves some important chemical principles that you understand, learn and apply if you want the top grade.

Checkpoint 2

Write an equation to suggest how urea (carbamide), $CO(NH_2)_2$, might react with nitrous acid.

amino acids

Azo-dyes

→ Diazotization is the formation of stable aqueous diazonium cations by reaction of a primary aromatic amine with nitrous acid at 0–5 °C:

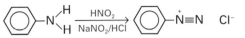

Phenylamine in HCl(aq)   Benzenediazonium chloride

The jargon

Nitrous acid is $NaNO_2$ in HCl(aq) to form HNO_2 *in situ*.

→ Benzenediazonium chloride is too unstable to be isolated but we use its solution in coupling reactions with phenols or aromatic amines to produce azo-dyes:

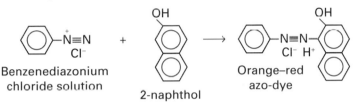

Benzenediazonium   2-naphthol   Orange–red
chloride solution   azo-dye

The jargon

A *coupling* reaction joins a diazonium compound and another molecule (usually to form an azo-dye).

Because diazonium compounds couple with phenols and amines it is very important during the preparation of benzenediazonium chloride

1   to keep the temperature below 10 °C so no phenol is formed
2   to diazotize all the phenylamine so none is left to couple

→ Bromine water reacts with a solution of phenylamine in HCl(aq) to give a whitish precipitate of 2,4,6-tribromophenylamine:

$$C_6H_5NH_2(aq) + 3Br_2(aq) \rightarrow C_6H_2Br_3NH_2(s) + 3HBr(aq)$$

Grade booster

There are often questions on amines and diazotization of phenylamine. If you want the top grade make sure you get the marks we give for the experimental conditions.

Checkpoint 3

Draw the structure of 2,4,6-tribromophenylamine.

Amino acids

→ Amino acids contain an amino group and a carboxyl group:

2-aminopropanoic acid NH_2 NH_2 3-aminopropanoic acid
α-aminopropanoic acid CH_3CHCO_2H $CH_2CH_2CO_2H$ β-aminopropanoic acid

α-amino acids

→ About 20 α-amino acids (2-aminoalkanoic acids) occur naturally and produce proteins essential to life.
→ The R group may be one of about 20 different groups:

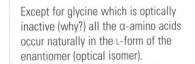

Don't forget!

Except for glycine which is optically inactive (why?) all the α-amino acids occur naturally in the L-form of the enantiomer (optical isomer).

side-group	neutral		acidic	basic
—R	= —H	—CH₃	—CH₂CO₂H	—CH₂(CH₂)₃NH₂
name	= glycine	alanine	glutamic acid	lysine
code	= gly	ala	glu	lys

Glycine, $CH_2NH_2CO_2H$, is the simplest α-amino acid and is typical in forming a solid with a crystal lattice of *zwitterions*: $^+NH_3CH_2CO_2^-$.

→ In alkali (pH > 7) the zwitterion becomes $NH_2CH_2CO_2^-$
   In acid (pH < 7) the zwitterion becomes $^+NH_3CH_2CO_2H$

The jargon

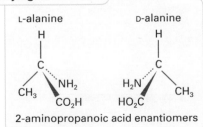

L-alanine   D-alanine

2-aminopropanoic acid enantiomers

Exam practice (5 minutes)   answers: page 190

Suggest why, during diazotization (a) phenylamine is in HCl(aq), (b) coupling agents like 2-naphthol are in alkaline solution, and (c) benzenediazonium ions are more stable than ethyldiazonium ions.

Compounds containing nitrogen: proteins and

Polyamides occur naturally as proteins. Proteins are built from long chains of amino acids linked by peptide bonds. Shorter chains are called polypeptides. Nylons are synthetic polyamides.

Proteins

Optical isomers

The jargon

Optical isomers (*enantiomers*) are non-superimposable mirror-image structures, e.g.

Letter specifies form Sign shows direction
of structure of optical rotation
L(+)-glutamic acid D(−)-glutamic acid

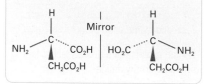

All the α-amino acids *except* glycine have chiral molecules with four different atoms or groups attached to a carbon atom, so they can show *optical isomerism*.

→ All naturally occurring amino acids from protein are L-forms.

An amino group in one molecule and a carboxyl group in another molecule can *in principle* lose a water molecule to form an amide link:

The jargon

Amide (peptide) link = —CO—NH—

$$NH_2-\underset{R}{\underset{|}{\overset{H}{\overset{|}{C}}}}-CO_2H + NH_2-\underset{R}{\underset{|}{\overset{H}{\overset{|}{C}}}}-CO_2H \xrightarrow{-H_2O} NH_2-\underset{R}{\underset{|}{\overset{H}{\overset{|}{C}}}}-\overset{O}{\overset{||}{C}}-\underset{}{\overset{H}{\overset{|}{N}}}-\underset{R}{\underset{|}{\overset{H}{\overset{|}{C}}}}-CO_2H \quad = a\ dipeptide$$

M_r of proteins ranges from 5×10^3 to 4×10^7. Polypeptides like oxytocin (causes uterine contraction in childbirth) are proteins with small relative molecular masses.

Action point

Make a table of amino acids and the structures of their side-chains. Group them into acidic, basic or neutral side-chains.

→ Proteins are natural polyamides with only one —CHR— between each amide (peptide) linkage:

$$NH_2-\underset{R_1}{\underset{|}{\overset{H}{\overset{|}{C}}}}-\overset{O}{\overset{||}{C}}-\underset{}{\overset{H}{\overset{|}{N}}}-\underset{R_2}{\underset{|}{\overset{H}{\overset{|}{C}}}}-\overset{O}{\overset{||}{C}}-\underset{}{\overset{H}{\overset{|}{N}}}-\underset{R_3}{\underset{|}{\overset{H}{\overset{|}{C}}}}-\overset{O}{\overset{||}{C}}-\underset{}{\overset{H}{\overset{|}{N}}}-\underset{R_4}{\underset{|}{\overset{H}{\overset{|}{C}}}}-\overset{O}{\overset{||}{C}}-OH$$

→ The R group may be any one of about 20 different groups in any order along the protein chain.
→ The *primary structure* of a protein is the sequence of amino acids in the chain, often shown by codes: GlyAlaLysGluGluSerMet. . . .
→ S—S bonds, hydrogen bonding and electrostatic attractions involving side groups cause chains to cross-link, coil into a helix and fold into pleats to produce the *secondary structure* of proteins.
→ The *tertiary structure* is the overall three-dimensional shape of (cross-linked, coiled and pleated) protein molecules.

The jargon

Enzymes are important biological catalysts.

Enzymes, antibodies, haemoglobin, casein, albumin and insulin are globular proteins soluble in water. Keratin and collagen (in hair and muscle) are fibrous proteins insoluble in water.

Proteins and chromatography

The jargon

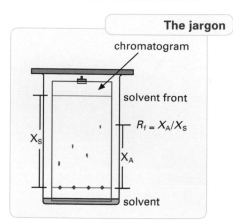
chromatogram

solvent front

$R_f = X_A/X_S$

X_S

X_A

solvent

→ Chromatography separates and identifies components in a mixture. We can hydrolyze a protein with HCl(aq) to a mixture of amino acids

$$\ldots -CHR-CO\!-\!\!-NH-CHR'- \ldots H_2O$$
$$\rightarrow \ldots -CHR-CO_2H \quad NH_2-CHR'- \ldots$$

and use *paper chromatography* to separate the mixture. We spray the paper with ninhydrin. On drying and warming, purple spots appear and the amino acids are identified by comparing their R_f value with data book values.

polyamides

Amides and polyamides

Amides

Amides have the general formula $RCONH_2$ and are derived from carboxylic acids RCOOH by replacing the OH with NH_2:

$$CH_3COOH + NH_3 \rightarrow CH_3COONH_4 \xrightarrow{\text{heat}} CH_3CONH_2 + H_2O$$
$$CH_3COCl + 2NH_3 \rightarrow CH_3CONH_2 + NH_4Cl$$

Polyamides

Proteins are *natural* polyamides and nylons are *synthetic* polyamides with repeating units joined by the peptide or amide link which can be

represented by ![amide link structures], —CO—NH- or -CONH-

In different nylons the group between each link will differ:

nylon-6–10
$$NH-(CH_2)_6-NH-CO-(CH_2)_8-CO-NH-(CH_2)_6-NH-CO-(CH_2)_8-CO-$$
nylon-6–8
$$NH-(CH_2)_6-NH-CO-(CH_2)_6-CO-NH-(CH_2)_6-NH-CO-(CH_2)_6-CO-$$
nylon-6
$$NH-(CH_2)_5-CO-NH-(CH_2)_5-CO-NH-(CH_2)_5-CO-$$

A sample of nylon-6–10 can be prepared in the laboratory at the interface of a solution of $NH_2-(CH_2)_6-NH_2$ in water and $ClCO-(CH_2)_8-COCl$ in tetrachloromethane. $NH_2-(CH_2)_8-NH_2$ in water and $ClCO-(CH_2)_8-COCl$

Nylon-6 is made industrially from a single monomer called caprolactam. Under suitable conditions this cyclic amide splits open and polymerizes.

The polymer is spun into fibres that have elasticity and high tensile strength because the long-chain molecules can coil and stretch.

Exam practice

answers: page 190

(a) Outline the preparation of phenylamine from nitrobenzene.

(b) State how phenylamine reacts with nitrous acid.

(c) Draw the isomers of 2-amino propanoic acid.

(d) State the effect of adding aqueous sodium hydroxide to (i) the ammonium salt of a carboxylic acid; (ii) a primary amine; (iii) an amide of a carboxylic acid.

(e) State and explain what you would expect to happen if nylon is refluxed with aqueous hydrochloric acid.

Checkpoint 1

Suggest an equation for the formation of ethanamide from ammonia and ethanoic anhydride.

Grade booster

Make sure you know
→ the difference between amides and amines
→ an example of a primary, secondary and tertiary amine
→ the relationship between α-amino acids and proteins
→ an example of an important polyamide
→ the differences between proteins and polyamides

Checkpoint 2

Write an equation to show the reaction of $ClOC-(CH_2)_8-COCl$ with $NH_2-(CH_2)_6-NH_2$ to show the formation of nylon-6–10.

Grade booster

Remember that marks are given for the reagents and reaction conditions used.

Watch out!

Kevlar is a polyamide that is quite different in its physical properties from the nylons.

Kevlar

It was used to strengthen the fuel tanks in the supersonic aircraft, Concorde. In 2008, Prof. Zhong Lin Wang reported in Prof. Zhong Lin Wang of Georgia Tech in Atlanta, Georgia, USA, reported that pairs of semi-conducting zinc oxide nanofibres, grown on Kevlar, can generate piezoelectricity when twisted, bent and moved. Multiple pairs could, in theory, be woven into fabric (for clothing, etc.) to produce electricity from its mechanical movement – so you could power your iPod as you walk.

Instrumental techniques: UV, visible and

Grade booster

Master these topics if you want to achieve the top grade. They help examiners pick out the best candidates.

Chemists have combined computer technology and spectroscopy into very powerful tools for analysing and identifying compounds. You need to understand the principles and be able to interpret simple simple spectra.

Ultraviolet and visible spectroscopy

We see colours because (electronic transitions in) compounds absorb light of different wavelengths from the visible region of the electromagnetic spectrum. Electrons move between d-orbitals in d-block compounds and within conjugated systems of π-electrons in organic *chromophoric* compounds. Our eyes cannot see UV light but many compounds absorb light in the UV region.

Checkpoint 1

Draw and label a line to represent the electromagnetic spectrum. Include radiowaves, microwaves, X-rays, ultraviolet, infrared and visible radiation in order of decreasing frequency.

The jargon

A *chromaphore* is an ion, molecule or group absorbing UV or visible light; e.g. −N=N− absorbs blue so azo-dyes are often orange-red.

Split-beam UV/visible spectroscopy

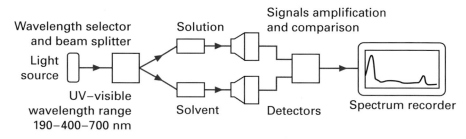

The instrument automatically compares the absorptions by the solution and pure solvent (both in identical glass or quartz cells) to get the absorbance by the solute only. The concentration of the solute in the solution is related to the absorbance by the Beer–Lambert law.

Checkpoint 2

(a) At which end of the visible spectrum (red–orange–yellow–green–blue–indigo–violet) is the maximum absorbance of phenolphthalein in basic solution?
(b) Sketch a possible absorption spectrum in the visible region for chlorophyll.

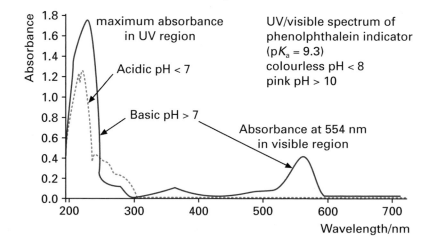

→ UV and visible spectroscopy is used as an analytical tool to study d-block metal ions and complexes, to determine structure and to study the kinetics of chemical reactions.

Grade booster

You need not know the detailed instrumentation and mode of operation of a UV/visible spectrometer but you need to know and understand the broad underlying principles of the technique if you want the top grade.

Infrared spectroscopy

→ Molecules can absorb IR radiation as *vibrational* energy that excites the natural bending and stretching of their bonds:

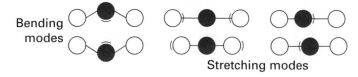

Bending modes

Stretching modes

IR spectroscopy

Infrared spectrometers are similar to UV/visible instruments but the sample must be in a cell made of crystalline sodium chloride because (unlike glass which it resembles) the salt does *not* absorb IR radiation. The instrument automatically

1 varies the IR frequency beamed into the sample
2 detects the radiation transmitted and
3 records the results as a spectrum of transmission against wavenumber

The absorption spectrum is like a molecular 'fingerprint' because the different functional groups in a molecule have different characteristic absorption frequencies. However, the spectra are complicated because these characteristic frequencies may change with the position of the functional group in the molecule.

→ Advantages of IR spectroscopy – we only need a very small sample for a good spectrum and we get results quickly.

Infrared spectrum of hexan-2-one

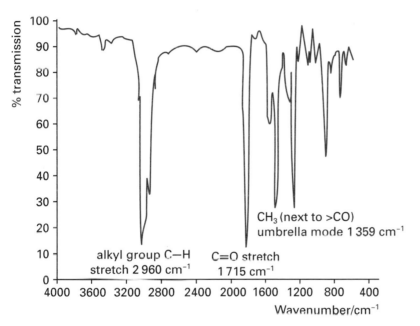

Examiner's secrets

You will always be given these frequencies in an exam. You won't have to remember them.

The jargon

For UV and visible spectroscopy we use wavelength in nm.
For IR spectroscopy we use wavenumber in cm^{-1}.
Wavenumber = 1/wavelength.

Checkpoint 3

Draw a diagram to describe the CH$_3$ umbrella mode of vibration.

Watch out!

See page 123 for an outline of atomic emission spectroscopy.

Watch out!

[iodine]/10^{-4} × mol dm^{-3} = 5.40 means the iodine concentration is 0.000 540 mol dm^{-3}

The jargon

PABA (para-aminobenzoic acid) = 4-aminobenzoic acid.

Exam practice   (6 minutes, 10 minutes)   answers: pages 190–191

1 A spectrometer was used to measure the colour of aqueous iodine and follow the change in concentration with time in a reaction mixture. Use the data to plot a suitable graph and find the order of reaction with respect to the iodine.

[iodine]/10^{-4} mol dm^{-3}	5.40	4.92	4.28	3.64	2.84	2.20
time/min	0.0	3.0	7.0	11.0	16.0	20.0

2 Discuss briefly how UV and IR spectroscopy might be used (a) to detect a trace impurity in a sample of PABA (an organic compound used in suntan creams) and (b) to identify the gases in a sample from a faulty car exhaust system.

Instrumental techniques: NMR and mass spectrometry

Mass spectrometers and NMR spectrometers are powerful tools for determining the structure of organic compounds. Both kinds of instrument use magnetic fields but in different ways.

NMR spectroscopy

The nuclear magnetic resonance instrument

1. surrounds the sample with an extremely powerful magnetic field
2. passes radiofrequency waves (from an oscillator) into the sample
3. scans by varying magnetic field strength *or* radiowave frequency
4. detects and records the results as an absorption spectrum

The technique works with the nuclei of ^1H, ^{13}C, ^{19}F and ^{31}P because they have an odd number of protons (or neutrons). A 14.1 T (tesla) magnetic field strength needs a radio frequency of 60 MHz for ^1H. The frequency of the radiowaves absorbed depends on the environment of the nucleus. Hydrogen atoms in an organic compound usually have different environments. Consequently, NMR spectroscopy is now one of the techniques most widely used by organic chemists.

→ Doctors now use NMR scanners on people as part of their diagnosis of diseases.

NMR low and high resolution spectrum for ethanol

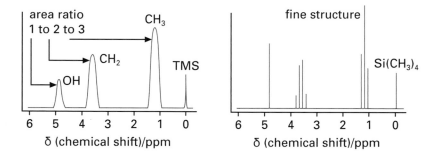

→ In low resolution NMR, the ratio of peak areas tells us the ratio of the numbers of each kind of proton (^1H) in the molecule.

We expect three integrated peaks in ethanol (there are three types of ^1H) with the areas in the ratio $3 : 2 : 1$ (in CH_3, CH_2, and OH).

→ Tetramethylsilane (TMS) is the standard to set the scale because its 12 H-atoms are equivalent and it gives only one peak.

We mix a small amount of TMS with the sample and the instrument generates a chemical shift scale. The chemical shift depends on the particular environment of the proton.

→ In high resolution NMR, peaks split when adjacent carbon atoms have H-atoms attached that are not equivalent.

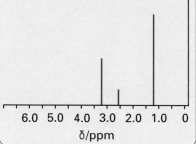

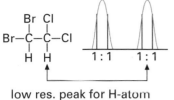 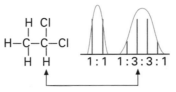

low res. peak for H-atom
splits into two high res.
peaks of equal intensity

low res. peak for H-atom
splits into four high res.
peaks of different intensity

If there are n equivalent hydrogen atoms on the C-atom next to the carbon carrying the ^1H, then the low resolution peak for that ^1H will split into $n + 1$ high resolution peaks whose relative intensities can be calculated from Pascal's triangle.

n	Relative intensities of peaks
1	1 1
2	1 2 1
3	1 3 3 1
4	1 4 6 4 1

Mass spectrometry

→ Instrument ionizes (and fragments) gas molecules into a beam of *positive* ions to be scanned by a varying magnetic field.
→ Signal strength (amplitude) of detected current is proportional to the abundance of ions of a given mass/charge ratio.

If some molecules do not fragment, the spectrum may show a peak for the molecular ion and give us a value for the relative molecular mass.

Incomplete mass spectrum of ethanol

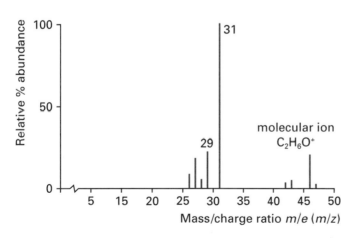

answers: page 191

Checkpoint

Suggest why mass spectrometers must be kept at a very low pressure.

The jargon

The *mass/charge ratio* is the mass of the ion divided by its positive charge. A *molecular ion* is a whole (unfragmented) molecule that has lost one electron.

Watch out!

Naturally occurring isotopes of C, H and O complicate high-resolution mass spectra and have to be taken into account.

Grade booster

If you want the top grade make sure that you understand the principles of mass spectrometry and get plenty of practice interpreting spectra.

Exam practice (6 minutes, 3 minutes)

1. For the mass spectrum of ethanol, suggest and explain (a) what causes the peaks at $m/e = 31$ and 29, (b) two possible peaks not shown, (c) why there is a very small peak at $m/e = 47$ and (d) what other type of spectrum might help to confirm ethanol as the compound giving this mass spectrum.

2. For methanol suggest and explain how many peaks would be in the low resolution NMR spectrum and the ratio of their areas.

Analytical techniques: chromatography

Chromatography is the separation of mixtures for analysis and/or purification by physical methods. There are a variety of physical methods but they all involve a mobile phase to carry the mixture over a stationary phase.

Grade booster

Russian botanist Mikhail Tsvet invented the first chromatographic technique in 1900 while investigating chlorophyll. Martin and Synge were awarded the Chemistry Nobel Prize In 1952 for their invention of partition chromatography.

Action point

Revise paper chromatography: see page 162

Thin layer chromatography (TLC)

This is similar to paper chromatography. In thin layer chromatography the stationary phase is a uniform thin layer of silica gel or alumina spread over the surface of a thin glass plate or plastic sheet.

Checkpoint 1

(a) What is the stationary phase In paper chromatography?
(b) What mobile phase could be used to analyse (i) chlorophyll and (ii) permanent black ink?

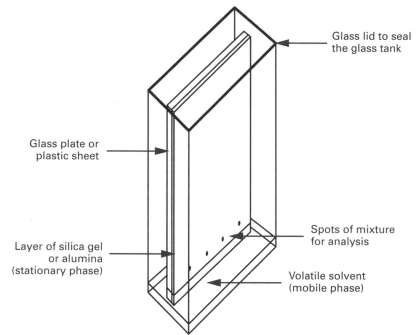

Glass lid to seal the glass tank

Glass plate or plastic sheet

Layer of silica gel or alumina (stationary phase)

Spots of mixture for analysis

Volatile solvent (mobile phase)

Checkpoint 2

(a) Why should the sample spots be as small as possible?
(b) Why should the solvent not be allowed to reach the top of the plate?
(c) What might be the effect on the chromatogram of not sealing the tank?

The layer of silica or alumina is spotted with a small samples of the mixtures. The plate is then suspended in a tall vessel with the spots just above the surface of the solvent. The vessel is covered so that the plate is surrounded by solvent vapour kept at its equilibrium pressure. As the solvent front rises up the plate, the components separate. When the solvent has reached a suitable height, the plate is removed and, where necessary, developed.

→ The **retardation factor** or R_f value is defined as

$$\frac{\text{Distance moved by component}}{\text{Distance moved by solvent front}}$$

It is easy to measure R_f values if the components are coloured but if they are not visible the plate may have to be developed with a reagent that makes the spots visible. In some cases the components may be seen under ultraviolet light.

→ Components may be identified by their measured RF values or by comparison with reference samples placed on the plate.

→ TLC needs only small samples and its sensitivity, speed and accuracy makes it suitable for use by forensic scientists

Gas Liquid Chromatography (GLC)

GLC is a highly sensitive way of separating organic mixtures. It may be used quantitatively and in conjunction with mass spectrometry. The mobile phase is an inert carrier as such as nitrogen, argon, helium etc. The stationary phase is an inert non-volatile oil supported upon solid particles packed in a long coiled tube (or column) contained within a thermostatically controlled oven.

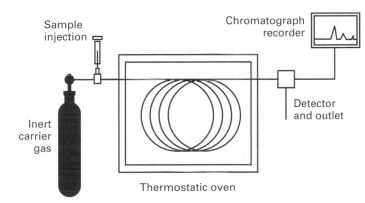

Jargon

Retention time is the time taken for a component to reach the GLC detector.

The components distribute themselves between the carrier gas and the oil in the column so some move more quickly than others do through the instrument. As each component of the injected mixture emerges from the column, it is detected, recorded and identified by its retention time. The amount of a component in the sample mixture may be estimated by integrating the area under a peak in the chromatograph.

Checkpoint 3

Describe a simple picture (in terms of the stationary oil phase, the mobile gas phase and molecules in the sample mixture) to explain why components have different retention times.

High Performance Liquid Chromatography (HPLC)

In HPLC the stationary phase is a column of small particles packed in a small diameter tube. The mobile phase is a liquid pumped under high pressure through the column. Excellent separation of complex mixtures is achieved by carefully selecting the stationary phase (chemical nature and particle size), mobile phase, column diameter and length, as well as the temperature and the pressure. Components may be identified and estimated as in GLC.

Watch out!

By using a **liquid** under high pressure, HPLC are shorter and separations are better than those in GLC.
When we use GLC and HPLC for analysis, we need only small samples. When we employ GLC and HPLC for preparation and purification, we use much larger amounts of materials.

Applications of GLC and HPLC

→ accurate identification and determination of drugs in body fluids, e.g. alcohol in blood.

→ separation, isolation and purification of chemicals.

→ quantitative analysis, e.g. determination of the concentration of a compound.

→ analysis in conjunction with mass spectrometry

Exam practice   answers: page 191

Explain how GLC can improve the function of a mass spectrometer.

Industrial chemistry: oil refining and petrochemicals

We depend on petroleum (*rock oil*) for energy and a vast range of organic chemicals. Most of this non-renewable resource is refined and used as fossil fuels. The rest is processed into the raw materials for the chemical industry.

Oil refining

→ *First stage*: distillation to separate crude oil into fractions:

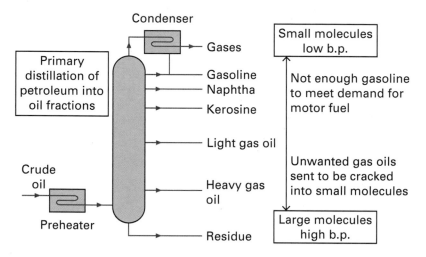

Grade booster

You don't need to learn distillation and cracking flow diagrams by heart but you should know and understand the principles and differences of the processes if you want the top grade.

Watch out!

Saturated alkanes crack to a mixture of alkanes, hydrogen and alkenes (or other unsaturated hydrocarbons).

The jargon

A *fluidized bed* is a fine powder behaving like a liquid when gas passes through.

→ *Second stage*: thermal, steam or catalytic cracking to break large molecules into small ones, e.g. $C_{14}H_{30} \rightarrow C_{10}H_{22} + 2C_2H_4$.
→ Cracking of petroleum fractions is a major source of ethene, propene and hydrogen for the chemical industry.
→ In catalytic crackers hot hydrocarbon vapour flows through fluidized beds of *heterogeneous catalysts*

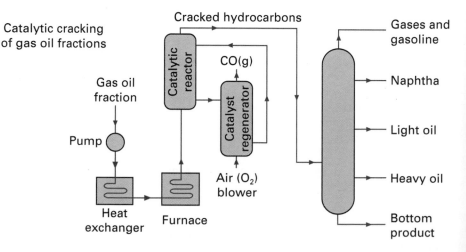

→ *Third stage*: isomerization and reforming.
→ Isomerization is the rearrangement of a molecule into an isomer: e.g. $CH_3CH_2CH_2CH_3 \rightarrow (CH_3)_3CH$ using Pt on Al_2O_3 catalyst.
→ Reforming is the conversion of a molecule into a different type: e.g. $CH_3(CH_2)_5CH_3 \rightarrow C_6H_5CH_3 + 4H_2$.

Chemicals from petroleum

The following is a selection of petrochemicals and related products you may well encounter in your A-level course.

Ethene

chloroethene: $CH_2=CH_2Cl$ (to make PVC); epoxyethane: CH_2-CH_2 with O bridge;

poly(ethene): $-(CH_2-CH_2)_n-$; ethyl benzene: $C_6H_5CH_2CH_3$;

ethane-1,2-diol: CH_2OHCH_2OH (antifreeze and to make polyesters).

Propene

propan-2-ol: $CH_3CH(OH)CH_3$; poly(propene): $-(CH_2-CH(CH_3))_n-$;

acrylonitrile: $CH_2=CH_2CN$ (to make acrylic plastics); cumene: $C_6H_5CH(CH_3)_2$

(to make phenol and propanone) – see below.

Butadiene

$CH_2=CH-CH=CH_2$ (to make synthetic rubber and other products).

Alicyclics

cyclohexane C_6H_{12} and cyclohexanol $C_6H_{11}OH$ (used to make

1,6-hexanedioic acid and caprolactam for nylon manufacture).

Aromatics

benzene: $C_6H_6 \rightarrow C_6H_{12}$ cyclohexane (to make nylon-6)

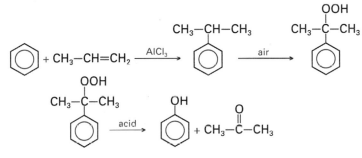

methylbenzene (toluene): $C_6H_5CH_3$ (to make TNT)

Azo-dyes

methyl orange

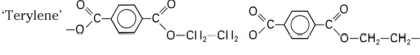

Polymers

poly(phenylethene)
polystyrene

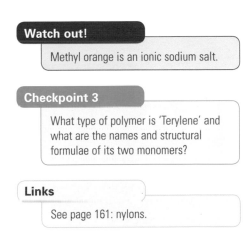

'Terylene'

Carbon black

fine particles of the element (in printing inks and tyre manufacture)

Nylon – 6

$- CO(CH_2)_4 \, CONH(CH_2)_6 \, NHCO(CH_2)_4 \, CONH(CH_2)_6 \, NH -$

Links

For uses of important organic feedstocks visit the following website if you want the top grade http://www.iupac.org/didac/Didac%20Eng/Didac02/Content/P12.htm

Checkpoint 1

State the type of polymer formed by acrylonitrile and draw the structure showing four repeating units.

Checkpoint 2

State and explain the type of reaction for the formation of cumene from benzene and propene.

Watch out!

Methyl orange is an ionic sodium salt.

Checkpoint 3

What type of polymer is 'Terylene' and what are the names and structural formulae of its two monomers?

Links

See page 161: nylons.

Exam practice (5 minutes)   answers: page 192

(a) In the catalytic cracking of oil suggest (i) why the catalyst needs to be regenerated, and (ii) what might occur in the catalyst regenerator.

(b) What type of molecule (class of compound) is represented by

(i) $CH_3(CH_2)_5CH_3$ and (ii) $C_6H_5CH_3$?

How Chemistry Works 1: data and patterns

There are many kinds of chemists (analytical, biochemical, industrial, inorganic, nuclear, organic, physical, theoretical, etc.) doing different things in different ways. Nevertheless, their work usually involves, in some way or another, collecting data, finding patterns, establishing principles, testing theories and publishing their work for evaluation.

Collecting data

Nowadays chemists rely heavily on instruments of increasing sensitivity and accuracy to collect and analyse qualitative and quantitative data. We can, for example, identify and measure trace constituents in samples from the earth's atmosphere or the moon's surface, analyse polar ice core samples to estimate historical global climate changes, extract and compare the DNA from tiny samples of blood, skin or other tissue, as well as detect and determine drug traces. Whatever methods we use (chromatography, mass spectrometry, IR & UV spectrophotometry, NMR, X-ray diffraction, etc.) and however accurate our data, we must always ensure that our results are consistent, reproducible and trustworthy. For that reason we publish our results for other chemists to check and verify.

Finding patterns

The Electrochemical Series

Metals have an ancient history. Copper, tin and their alloy, bronze, were known and used more than 5000 years ago. The first battery (using iron and copper) may have been made more than 2000 years ago. However, the first recorded electrochemical effect refers to 1786 when Luigi Galvani touched a dissected frog's leg simultaneously with two different metals and made it twitch.

In 1800 Alessandro Volta published details of his battery made from a pile of triple layers of silver, cloth soaked in salt, and zinc. The more triple layers (Ag::Zn,Ag::Zn,Ag::Zn, etc) the higher the voltage and the great the current. Volta also experimented with different metals and introduced the idea of an electrochemical series. In 1807 Sir Humphry Davy isolated sodium and potassium by electrolysis of their molten hydroxides. In 1808 Jöns Jakob Berzelius discovered lithium.

Today we have an electrochemical series for metals and non-metals based on standard electrode potentials.

The Periodic Table

You should be familiar with the Periodic Table but you may not know its history and you may not realise new developments are still being proposed. The Periodic Table is the outcome of our search for patterns in the collection of data that began in 1649 when Hennig Brand discovered the chemical element phosphorus. 62 more elements had been discovered by 1869.

→ In 1817 Johann Dobereiner arranged elements into triads according their atomic weights.

→ In 1862 A.E.Beguyer de Chancourtois, a French geologist, arranged elements (and unfortunately some compounds) on a cylinder and saw that properties repeated every seven elements and predicted the stoichiometry of several metallic oxides.

→ In 1864 John Newlands published a periodic table and his Law of Octaves proposing that any given element will exhibit analogous behavior to the eighth element following it in the table.

→ In 1864 and 1868 the German chemist, Lothar Meyer, produced an abbreviated (published in his textbook) and an extended version (given to a friend) of his table showing periodic changes of valence with atomic volumes.

→ In 1869 the Russian chemist, Dmitri Mendeleev, working independently of Meyer, published his table one year before Meyer's appeared in 1870.

→ In 1895 Lord Rayleigh reported his discovery of argon.

→ In 1898, William Ramsey proposed a new (zero) group for argon (between chlorine and potassium) in the periodic table with helium as the first member and accurately predicted the discovery and properties neon

→ In 1911 and 1913 separate publications by Ernest Rutherford, A. van den Broek and Henry Moseley led to the charge on the nucleus, atomic number (Z), isotopes and the need to place elements in the Periodic Table by atomic number.

→ In 1940 Glenn Seaborg discovered plutonium. Thereafter he found all the transuranic elements from 94 to 102 and rearranged the Periodic Table by placing the actinide series below the lanthanide series.

Amongst the papers (of a international conference held in 2003) published under the title "The Periodic Table: Into the 21st Century" Maurice Kibler considers the possibility of classifying not only the chemical elements but also the subatomic and subnuclear particles with a symmetry SO(4,2)xSU(2) table and Jerry Dias describes a periodic table (of benzenoid hydrocarbons to fullerene carbons) as an aid to systematizing information about these organic pollutants in the environment.

Homologous Series

You should already know that chemists can handle and make sense of the increasingly vast number of organic compounds by arranging them into classes based on patterns in their structure and properties of their molecules and functional groups.

Exam practice	answers: page 192

a) Plot the following numerical values (determined from Boyle's original publication) of reciprocal volume of air (1/V) on the y-axis against pressure (P) on the x-axis.

P	48	46	44	42	40	38	36	34	32	30	28	26
1/V	3.43	3.27	3.13	2.99	2.83	2.70	2.54	2.40	2.26	2.12	1.99	1.84

P	24	23	22	21	20	19	18	17	16	15	14	13	12
1/V	1.70	1.63	1.56	1.49	1.41	1.35	1.28	1.21	1.14	1.07	1.00	0.93	0.85

(b) Discuss briefly the suggestion that Boyle's published data may not be trustworthy.

The jargon

In the time of Lothar Meyer and Dmitri Mendeleev
Atomic weight – was the mass in grammes of the ratio of the mass of an atom to the mass of a hydrogen atom.
Atomic volume – was the volume of the atomic weight of an element in the solid state.

Grade booster

Top candidates would know that Mendeleev left gaps in his table and predicted the existence of three elements (eka-boron, eka-silicon that were later discovered and whose properties agreed quite well with his predictions. He actually predicted 10 new elements altogether three of which (at. wt. 45, 146 & 175) do not exist.

Checkpoint 2

What is the modern name for eka aluminium, eka-boron and eka-silicon?

Checkpoint 3

How did the work of Niels Bohr (into atomic structure) and G.N. Lewis (into electronic structure) provided an explanation for the periodic table and the periodic law?

Watch out!

Glenn Seaborg was awarded the Nobel Prize in 1951 and element 106 was named seaborgium (Sg) in his honor.

Action point

Check out the following webpage for a more detailed history of the periodic table http://www.wou.edu/las/physci/ch412/perhist.htm

Action point

Check out the following webpage for details of Robert Boyle's experiments that led to the gas law pV = constant for a fixed mass of gas at a fixed temperature. http://dbhs.wvusd.k12.ca.us/webdocs/GasLaw/Gas-Boyle-Data.html

How Chemistry Works 2: principles, theories

The work of chemists is perhaps best recognised by the practical applications that are developed and managed for the benefit of society. Unfortunately these many benefits are often forgotten and overshadowed by the deficit to society of industrial accidents, environmental pollution and global warming, all too often laid at the door of chemistry.

Establishing principles and theories

You should already know some important principles of kinetics (what factors affect the rate of a reaction) and energetics (what determines a reaction's feasibility). The rate laws of various orders (zero, first, second, etc.) of reaction are accepted as having been established experimentally. The (first, second and third) laws of thermodynamics are accepted as having not been disproved (e.g. nobody has been able to construct a perpetual motion machine).

You should be aware that Niels Bohr proposed his planetary model of the atom on his mathematical analysis and interpretation of atomic spectra. Improved spectroscopy, ionisation energy data and advanced mathematical techniques led to the current principle of atomic and molecular orbitals and the rules governing the formation of chemical bonds.

You should also appreciate the important principle that the chemical properties of organic compounds are governed by their molecular structure and functional groups as well as by the types of reagent and the reaction mechanisms.

Testing theories

We generate a theory not only as a basis for the data we collect and patterns we observe but also as a guide to collecting further data in part as a test of our theory. For example, if a key reaction in the production of iron from its ore is reversible $Fe_2O_3(s) + 3CO(g) \rightleftharpoons 2Fe(l) + 3CO_2(g)$, then the percentage reduction of iron oxide would be governed by the law of chemical equilibrium. Early attempts to use up all the carbon monoxide (by building taller furnaces) failed and were abandoned when the thermodynamics of the process was understood.

Implementing applications

The discovery in 1936 in an ancient tomb near Baghdad of a clay pot, sealed with pitch, containing an iron rod surround by copper sheet suggests that over 2000 years ago Parthians may have electroplated gold onto silver artefacts. These ancient Iranians may even have coupled their batteries in series to achieve a higher voltage and current but their application of electrochemistry would probably not have been based on any theory.

A great number of chemical applications, predating modern theories, were based on collected data and observed patterns. Our ancestors knew how to extract metals and make alloys, produce alcohol and vinegar, make soap, extract dyes and colour cloth. Chemistry came into its own with the advent of theories to explain the chemical data and patterns they exhibited.

Watch out!

Rate equations must be determined experimentally and cannot be derived from a chemical equation.

Links

Find out how experimentally determined rate equations suggest mechanisms for the hydrolysis of C_4H_9Br and $(CH_3)_3CCl$. See page 149.

The jargon

To distinguish absorption and emission spectra
Visit http://csep10.phys.utk.edu/astr162/lect/light/absorption.html

Watch out!

Our ancestors discovered and performed a great deal of practical work long before chemistry became the modern mathematical and theoretical science it is today.

Action point

The word alkali is more than a thousand years old. It comes from the arabic *al kali* meaning *the burnt ashes*. Find out why.

and applications

Recent and ongoing developments

Carbon Nanotubes

In 1991 during experiments with fullerene, C_{60}, other carbon structures were discovered including carbon tubes made up of rolled up carbon hexagons.

 This sketch of part of a nanotube tube shows its extended graphite-like layer structure. These tubes are extremely fine in diameter (about 10 000 times thinner than a human hair) and conduct electricity like metals in some cases and like transistors in other cases. Researchers have now grown metal crystals inside these nanotubes so that they might be used for connections in miniaturised electronic circuitry too small for convential wires.

Fuel Cells

The fuel cell has come a long way since the German scientist, Christian Friedrich Schönbein, published his discovery of its principle in 1839 and the Welsh scientist, Sir William Robert Grove, had developed the first one by 1843. A fuel cell converts a reductant (a fuel like hydrogen or alcohol) at the anode and an oxidant (air or oxygen) at the cathode to produce electricity. A typical cell gives an emf of about 0.6 V but cells give *in series* higher voltages and *in parallel* higher current densities in a **fuel cell stack**.

In this simplified fuel cell, hydrogen ionises at the platinum catalyst anode into protons and electrons: $H_2(g) \rightarrow 2H^+(aq) + 2e^-$

The polymer/electrolyte membrane allows only the passage of positive ions to the cathode, forcing the electrons to the cathode via an external circuit. This means that the electrons power the external electrical device. At the cathode, oxygen, protons and electrons form the waste product of the device which is water: $O_2(g) + 2H^+ + 4e^- \rightarrow 2H_2O(l)$.

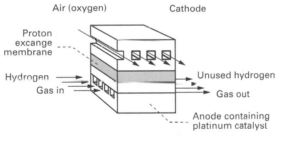

Air (oxygen) Cathode

Proton excange membrane

Hydrogen Gas in

Unused hydrogen Gas out

Anode containing platinum catalyst

Advantages of hydrogen fuel cells

→ Clean technology – water the waste product.
→ High efficiency and quiet operation – no moving parts.
→ Fairly simple construction.

Disadvantages of hydrogen fuel cells

→ Expensive and hydrogen storage poses problems.
→ Present cells last for an uneconomically short time.
→ Energy to produce the hydrogen exceeds the energy produced by the cell

Exam practice
answers: page 192

(a) State **three** advantages and **three** disadvantages of a motor vehicle running on fuel cells working on hydrogen gas.

(b) Suggest why carbon nanotubes may be more suitable for making connections in computers of the future.

How Chemistry Works 3: resources and society

In response to society's increasing demands for their products, chemical industries have become major users of energy and the earth's finite resources. At the same time the chemical industry has come under increasing governmental, legal and political pressure to economise, to increase efficiency and to minimise its impact upon the environment.

Energy and finite resources

Manufacture of aluminium

Since the first large-scale production plant opened in Pittsburgh in 1888, the process has been refined and improved. Other processes (ASP, drained-cell technology, inert anode, etc.) have also been used with varying success. However, they all start with Bauxite and finish with molten aluminium so that the overall energy requirements remain substantially the same and carbon oxide is one of the by-products. Australia, currently the world's largest producer of Bauxite, has an estimated reserve of around 7.4 billion tonnes. World reserves are estimated at around 34 billion tonnes. The supply could run out in the next 100 years.

Aluminium recycling

Aluminium drinks cans appeared in the USA in the early 1960s. A few years later, some aluminium manufacturers started buy-back recycling programs to recover the metal. Why? Recycling the metal is equivalent to a 95% saving of energy (expended in mining, refining and reducing the ore to the metal) and a 95% decrease in undesirable emissions into the environment. Unfortunately, in the USA less than 50% of the drink cans are recycled compared to around 85% of the cans in countries like Brazil, Japan, Sweden and Switzerland. In the USA an estimated 909 000 tonnes of aluminium metal drinks cans have been buried in landfills every year.

Manufacture of ammonia

Ammonia is manufactured (by direct combination of nitrogen and hydrogen) on the large scale as a heavy chemical required for the production of a range of other chemicals and materials including fertilisers and explosives. Our atmosphere is always the source of nitrogen. In 1910, in the first commercialised Haber-Bosch process the hydrogen came from the reaction(s) of steam with hydrocarbons (coal, natural gas, etc.). The production of an estimated 100 millions tons of ammonia-based fertiliser is believed to consume 0.75% of the world's energy supply and 3.4% of the world's natural gas. Other sources of hydrogen have been employed with varying success. For example, the electrolysis of water is extremely energy intensive and restricted to plants with access to low cost hydro-electricity.

Improving catalysts and lowering pressure

In 1992, the Kellog Advanced Ammonia Process (KAAP) was introduced using a ruthenium catalyst supported on active carbon. It operated at a pressure of 40 atm and not 100–250 atm and at a temperature around 350°C and not 450°C. In 2000, Danish scientists produced ruthenium catalysts with barium promoters that are 2.5 times more active than ruthenium on active carbon. They have also produced new, cheaper catalysts involving

The jargon

The *Hall-Héroult* process was discovered in 1886 independently by Charles Hall in America and Paul Héroult in France. In 1888, Hall opened the first large-scale aluminium production plant in Pittsburgh. *Bauxite* was named after Les Baux-de-Provence in Southern France, where geologist Pierre Berthier discovered it in 1821.

Action point

Revise the chemistry of aluminium extraction: see page 116.

Checkpoint 1

Suggest why the production of much thinner aluminium drinks cans resulted in a decrease in the number of cans recycled.

Watch out!

Is the energy expended in producing synthetic fertilizers via ammonia less than the energy we get from the food produced?

Action point

Look up the composition of a balanced synthetic fertilizer.

ternary nitrides (e.g. Fe_3Mo_3N, Co_3Mo_3N and Ni_2Mo_3N) and caesium promoters. In the past forty years, (a) the fuel efficiency of carbon dioxide removal during the production process has improved from 50 000 down to 10 000 kcal/kg-mol of CO_2 and (b) overall energy consumption has fallen from 42 to 30 billion BTU of natural gas per tonne of ammonia, with some new plants recovering more than 1 million BTU per tonne in the generation of electricity from heat that would otherwise be wasted.

The petrochemical industry

Many thousands of years ago our ancestors probably discovered oil or bitumen seeping from the ground and used it for fuel. More than 4000 years ago asphalt was used in building walls and towers in Babylon. Around 2000 years ago oil wells 800 ft deep were drilled in China. By the 8th century the Middle East petroleum industry was providing tar to line the streets of Baghdad. In 1858 the first commercial oil well was drilled in Ontario, Canada and in 1859 the petroleum industry began in the USA when Drake drilled his oil well in Pennsylvania. In 1967 the first oil sands surface mine started in Alberta. Today, the oil and gas industry is a multi-national business with a major impact on the gobal economy. About 85% of the petroleum ends up as fuels and 15% is processed into raw materials for the chemical industry.

Oil extraction

The Athabasca oil sands in Alberta contain a conservatively estimated 21.2 billion cubic metres of bitumen – second only to the oil reserves in Saudi Arabia. In 1883, C. Hoffman, a Canadian surveyor, reported that he easily separated bitumen from the oil sand using water. In 1926, Dr. Karl Clark at the University of Alberta perfected a commercially viable steam separation process now used for the surface mined bitumen. For bitumen situated 75 metres or more below the surface, cyclic steam stimulation (CSS) and steam-assisted gravity drainage (SAGD) in situ recovery methods have been developed involving thermal, solvent or carbon dioxide injection through vertical or horizontal wells.

Extraction processes use energy – about equal to that of 1000 cubic feet of natural gas for every barrel of oil (equivalent to about 6000 cu.ft.). At present the oil industry uses about 4% of Western Canada Sedimentary Basin natural gas production. As gas reserves are depleted, this fuel might be generated by gasification of bitumen in a simlar way to its convertion into synthetic crude oil and natural gas.

As the increasingly sophisticated search for oil and gas deposits continues, alternative renewable fuels are being considered. For example, alcohol (from the fermentation of cane sugar) has been used in Brazil as automotive fuel.

Action point

Find out how many joules in a BTU and what the three letters stand for. To do this visit http://www.simetric.co.uk/sibtu.htm

Jargon

Oil (tar) sands are deposits (often accessible by surface mining) of crude bitumen mixed with clay minerals, sand and water.

Checkpoint 2

Explain how water could separate oil from oil sand.

Action point

Visit http://en.wikipedia.org/wiki/Athabasca_Tar_Sands

Action point

Compile a list of the advantages and disadvantages of dealing with non-biodegradable plastics in domestic waste by incineration, landfill and recycling.

Exam practice   answers: page 192

(a) If in any one year 22 million tons of aluminium are manufactured and 7 million tons are recycled, estimate the total value of the metal burried in landfills over the past ten years. Assume the market price of aluminium to be £1250 per ton.

(b) Suggest and explain two obstacles to recovering aluminium metal from landfills.

How Chemistry Works 4: environmental and

The Jargon

The atmosphere above the earth's surface:

mesosphere   up to 50 km
stratosphere   up to 40 km
tropopause   narrow region between
troposphere   up to 10 km
ozone – O_3, an allotrope of oxygen
Dobson unit (DU) – is a measure of the amount of ozone above a point on the earth's surface and is named after G.M.B. Dobson, a British meteorologist who built the first ozone spectrophotometer in 1931.

Action point

Find out more about Dobson at http://www.theozonehole.com/dobsonunit.htm

The Jargon

CFCs – chlorofluorocarbon compounds (now banned) once used in foam manufacture, as aerosol propellants and in air conditioning/refrigeration units.

Action point

Find out about HCFCs and why they are safer than CFCs.

Watch out!

The chemical ozone is a greenhouse gas that could contribute to global warming but this should not be confused with depletion of the ozone layer.

Checkpoint

Determine the balanced equation for the overall reaction of the following catalytic cycle.
$ClO + ClO \rightarrow Cl_2O_2$
$Cl_2O_2 + hv \rightarrow Cl + ClO_2$
$ClO_2 \rightarrow Cl + O_2$
$2Cl + 2O_3 \rightarrow 2ClO + 2O_2$

Watch out!

The proposed reaction mechanisms to account for the depletion of the ozone layer are quite involved but not difficult for you to follow. Visit the University of Cambridge Centre for Atmospheric Science at the award-winning website: http://www.atm.ch.cam.ac.uk/tour/index.html

Through the work and technical expertise of chemists society has become increasingly aware of the damage human activities may inflict upon the environment. Green chemistry is broadly concerned with finding ways to reduce or eliminate such damage. The problems are complex and the solutions are rarely as simple as everyone would like.

The ozone layer

The facts

Ozone in the stratosphere is detected and measured (by spectrophotometers on the ground at Antarctica research stations, in aircraft flying scientific missions and on satelites) by the ozone's absorption of bands of ultraviolet (UV) radiation from back-scattered sunlight. The satellite-borne total ozone mapping spectrometer (TOMS) obtains a global picture of ozone levels.

The stratospheric ozone layer absorbs harmful UV radiation responsible for skin cancer, etc. Since the early 1970s when the work of Rowland and Molina (Nobel Prize winners in 1995) showed that CFCs could destroy ozone, the layer has been depleting and the breakdown products of CFCs have been detected in the stratosphere. The loss of ozone in mid-latitudes has been less and slower than the rapid and almost total loss of ozone over Antarctica at certain altitudes.

The theory

The ozone layer is thought to result from a balance of reactions that can occur at different rates at different altitudes. UV light dissociates oxygen molecules into radicals that react with other oxygen molecules to form ozone: (i) $O_2 + hv \rightarrow O + O$ then (ii) $O + O_2 \rightarrow O_3$. Any ozone dissociated by UV light: (iii) $O_3 + hv \rightarrow O_2 + O$ can reform (iv) $O_2 + O \rightarrow O_3$. Oxygen radicals can destroy ozone $O + O_3 \rightarrow O_2 + O_2$ but this reaction seems too slow to account for the depletion of the ozone layer. A faster reaction occurs with chlorine radicals (from CFCs), e.g.: $Cl + O_3 \rightarrow ClO + O_2$.

A key step in the theory is the production of Cl radicals by the UV dissociation of dichlorine peroxide: $Cl_2O_2 + hv \rightarrow Cl + ClO_2$. However, recent data collected by NASA chemists in California show this photolysis to be slower than previously thought. Consequently, around 60% of the observed ozone depletion at the poles must be due to an unknown mechanism. Researchers are now looking for an alternative theory.

Global warming

The facts

Carbon dioxide, dinitrogen oxide, ozone, methane and water are the principal greenhouse gases so-called because they contribute to the warming of the earth's surface by their ability to absorb and emit infra-red radiation. Recent and current levels of gases in the atmosphere have been measured by various methods including the use of radiosondes – small, expendable instruments carried by helium filled balloons that rise at about $300\,m\,min^{-1}$ to a height of around 35 km. Sensors linked to a battery powered radio transmitter on the radiosonde send measurements to a sensitive receiver on the ground. Since the 1980s, paleoclimatologists have used ice core geochemistry to determine Earth's climatic past and the effect of anthropological activities. Ice cores

global concerns

records provide the most detailed historical record available of greenhouse gas concentrations, dating back to times before humans began to influence the environment. Gas chromatographs with electron capture detectors measure gas concentration of air trapped in ice cores. The chronological uncertainties vary, for example, from 200 to several thousand years. Electroconductivity is used to identify interesting regions in a core and mass spectrometry is used to measure light, stable isotopes such as deuterium and ^{18}O. Core data indicates that atmospheric CO_2 level was around 260–280 ppm for 10 thousand years but over the last 240 years it rose by about 100 ppm – half that increase occurring in the past 40 years.

The theory
There are plenty of 'theories' (not all soundly scientific) and computer generated, mathematical models predicting serious future climate changes.

Your carbon footprint
You and your activities put (directly or indirectly) into the atmosphere greenhouse gases the total amount of which can be expressed as a CO_2 equivalent in kilogrammes or tonnes. You can estimate your 'carbon footprint' from the energy you use in the home (electricity, gas, etc.) and in travel (by car, bus, aeroplane, etc.) with an online calculator using a Life Cycle Assessment (LCA) method or one restricted to the directly attributable emissions from burning fossil fuels. These calculators rest on implicit assumptions and estimates that you should consider. For example, the consumption of 1 litre of propane is taken to be equivalent to 1.521 kg of carbon dioxide.

Theory, facts and practice
In trying to understand how scientists work you need to appreciate how we gather our facts (often more difficult outside than inside the laboratory), how we use those facts to establish (or refute) our theories and how we proceed on the basis of our theories. For example, where we explore for oil and how much we expect to find may depend upon our theory of how it formed. The conventional *biogenic* theory holds that petroleum was formed from dead prehistoric animals and plants over geological time by the anaerobic decay under raised pressure and temperature within certain rock formations. The controversial *abiogenic* theory holds that deep in the earth's mantle microbial reactions generate (from carbon, carbon dioxide, carbonates and methane) hydrocarbons that migrate into the earth's crust to be trapped under impermeable rock.

> **Jargon**
>
> *anthropological* activities – what human beings do and cause

> **Watch out!**
>
> There are sources of error in reconstructing climate history using ice core data, e.g., in determining age of the ice and correcting for gas exchange in snow prior to trapping of gas within bubbles enclosed in the compressed ice.

> **Action point**
>
> Estimate your carbon footprint online http://www.safeclimate.net/calculator

> **Watch out!**
>
> *Abiogenic* theory – proposed in Russia as early as 1757, revived and developed there in 1950–60 and supported by Professor Thomas Gold.

> **Action point**
>
> Make and compare a list of the scientific evidence (not personal opinions) in support of the biogenic and abiogenic theories of oil formation.

Exam practice (8 minutes)   answer: pages 192–193

(a) Write a balanced equation for the complete combustion of propane and calculate the atom economies of the carbon dioxide.

(b) Calculate the mass of carbon dioxide produced by the complete combustion of 1 litre of liquid propane (r.m.m 44.08; density 0.585 g cm^{-3}).

(c) Compare your result in (b) with the carbon footprint figure of 1.519 kg given for 1 litre of propane by the World Resources Center safe climate calculator and comment upon any difference.

How Chemistry Works 5: air, earth, fire and water

Since the discovery of fire, the advent of farming and the invention of the combustion engine, societies have been affecting the air, the earth and the water, often beneficially but all too often detrimentally and unknowingly.

Gas capture, storage and use

Carbon dioxide

For long-term storage, researchers internationally are evaluating geologic sequestration in which shale or porous rock adsorbs carbon dioxide (captured from industrial plants – power stations, cement production, etc.) and desorbs natural gas or oil. Since the year 2000, when the Saskatchewan CO_2 enhanced oil recovery project (designed to recover an incremental 130 million barrels from the Weyburn field) began, more than 1.9 billion cubic metres have been successfully sequestered and oil production incremented. Investigations in the USA indicate that 6.2 billion tons of CO_2 could be sequestered in the Big Sandy Gas Field area of eastern Kentucky alone. To minimise leakage to the surface, the sequestration should be beneath a cap of impermeable rock. Scotland plans to bury CO_2 (captured from its coal-burning power stations) deep beneath the sea bed. Edinburgh University researchers will explore the Firth of Forth for suitable sites.

Methane

Isotopic analysis for ^{13}C shows that about 20 per cent of the methane in the atmosphere was formed long ago and probably leaking from sub-oceanic rocks, coal seams, melting permafrost and other natural deposits. Since 1950 atmospheric methane concentration has been increasing at an estimated 1% per year – about four times the rate of increase of carbon dioxide. Annually 500 million tonnes of CH_4 could come from sources that include bogs and marshes, ~1.5 million km^2 of paddie fields, putrefying waste tips, intentional burning of forest and grassland and unintentional pipeline leaks. It has also been estimated that an annual 250 million tonnes world wide could come from cattle and termites.

→ methane is a serious 'greenhouse gas' on a par with carbon dioxide.
→ methane is a biofuel produced synthetically and from fermenters.
→ methane produces the least carbon dioxide per unit of energy of any fuel.

In 2006 there were more than 5 million natural gas vehicles (NGV) worldwide running on compressed (CNG) or liquified (LNG) natural gas. More NVGs are being made and more countries are adopting them.

Catalytic converters

Motor vehicles exhaust nitrogen and water together with carbon dioxide, carbon monoxide, oxides of nitrogen (NO_x) and unburnt fuel. To minimise undesirable emissions, sensors and computer control systems automatically adjust the fuel/air mixture to give optimum performance and keep the air-to-fuel ratio at or near the stoichiometric point for complete combustion. Three-way converters, consisting of a ceramic honeycomb structure (for optimum surface area) coated with platinum, rhodium and/or palladium metal, are used to catalyse the reduction of NO_x emissions to nitrogen and the oxidation of CO and unburnt fuel to CO_2 and H_2O.

Checkpoint 1

(a) Write an equation for the catalytic generation of methane from carbon monoxide and hydrogen.
(b) Explain why methane would be expected to produce per unit energy the least amount of carbon dioxide of any fuel

Hydrogen

The annual global industrial production of hydrogen is large (~50 million tonnes in 2004) and increasing at about 10% per year. Roughly half is used in the Haber process and half used in hydrocracking of heavy petroleum into lighter fractions for fuels. Hydrogen is a 'green' fuel and the subject of intensive research to replace our hydrocarbon economy.

Hydrogen can be either burnt directly in an internal combustion engine or used to generate electricity in a fuel cell. Mercedes-Benz, for example, has been testing its Citaro hydrogen fuel cell buses around the world in various cities, including London. Fuel cell vehicles (FCVs) can be more efficient and generate less pollution than conventional vehicles. Onboard storage of hydrogen presents a challenge. It may be stored as a liquid (at very low temperature), a compressed gas (at high pressure), an absorbed gas or a metal hydride. Organo-metallic coordination polymers (MOFs) and carbon nanotubes are potential physical absorbers for hydrogen.

Water, water everywhere

Fluoridation of the water supply

Depending upon the source and treatment method, your tap water may contain a variety of minerals and traces of the sterilising agent (usually chlorine) that local authorities use to remove harmful bacteria and supply drinking water which, in the UK, will certainly be as safe (if not safer) and far less expensive than (probably the same) water sold in plastic bottles over the counter. Some authorities 'fluoridate' the drinking water supply (particularly if little or no fluoride ions are present naturally) to prevent dental caries (tooth decay) especially in young children. The authorities' chemists ensure the aqueous fluoride ion is at the correct low concentration because at high concentrations there may be a risk of causing bone cancer and caries itself.

→ Fluoridation is the process of raising the concentration of aqueous fluoride ion in the water supply to the optimum level for dental health
→ The British Medical Association (BMA) found no convincing evidence of fluoridation being a risk to human health.

Nitrates

Rain may contain nitric acid formed in the atmosphere during lightning storms, etc. Other nitrates enter the soil by anaerobic plant decay, bacterial decomposition of excreta and other nitrogen cycle processes. The US Environmental Protection Agency recommends 45 ppm as the nitrate level not to be exceeded in potable water to minimise risk of blue baby syndrome – a condition which, though rarely fatal, poses a serious threat especially to infants of less than six months old. Recent research has linked (but not confirmed) other harmful effects of aqueous nitrate. Concern is growing over the rising levels of nitrate in ground and surface waters caused mainly by the widespread use of nitrogenous fertilisers (ammonium nitrate, ammonium sulphate, urea) and even manure.

Exam practice   answers: page 193

Describe the flame test to identify sodium and calcium ions in water samples and explain the theory behind the test.

How Chemistry Works 6: the green crystal ball

The Jargon

Ionic liquids – sometimes called designer solvents – are 'salts' consisting of a (usually bulky) organic cation and an inorganic (sometime omplex) anion. For example,

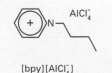

[bpy][AlCl$_4^-$]

RTILs are ionic liquids with m.pts. around *Room Temperature*.

Grade booster

For the top grade you should know the formula of the anion AlCl$_4^-$ but you will not be expected to know the names and structures of the corresponding organic cations in ionic liquids.

Watch out!

Sodium chloride above its melting point (800°C) would be an ionic liquid but not a useful solvent!

The Jargon

Supercritical liquid – a substance (usually under pressure) above its critical point, i.e. the temperature above which it will maintain its gaseous state and not turn into liquid.

Watch out!

Although ionic liquids have been termed *green solvents* (for not being detrimental to the ozone layer) some are more toxic (e.g. to aquatic life) than some VOCs. Balancing zero VOC emissions and avoiding spills into waterways should be a priority. Some are also combustible and need careful handling.

Checkpoint 1

(a) Write the systematic name for glycerol.
(b) State a use for trinitroglycerol.
(c) The –R groups in oils usually contain >C=C< bonds. State the effect on the physical nature of such oils when treated with gaseous hydrogen and a nickel catalyst.

In contrast to astronomers primarily concerned with the galactic distant past, chemists are focussed on the earth's present and future. Amazing developments have taken place in the short history of modern chemistry including analysis on a microchip and computer-aided drug design.

New methods and materials

Ionic liquids and supercritical fluids

The Montreal Protocol, an international agreement to control substances, such as volatile organic compounds (VOCs), that deplete the ozone layer, was signed in 1985 and has been updated at various times since. Ionic liquids are beginning to replace many VOCs as solvents and reaction media because of their negligible vapour pressure.

→ Ionic liquids do not evaporate even when exposed to vacuum.

Estimated at 700 billion tons, cellulose is our most widespread renewable organic chemical but less than 1% is used as a feedstock for lack of suitable solvent. The current cellulose paper and fibre production uses huge volumes of (disposable) water and volatile toxic solvents such carbon disulphide. According to research in Germany and the USA, ionic liquids serve as a real solvent for cellulose which can be easily released (in the form of fibres, etc.) from the solution with almost no loss the solvent (the ionic liquid) which can be recycled.

According to research in Britain, bimetallic palladium-ruthenium nanoparticles can be prepared, by the reduction of their simple oxides in (reverse micelles in) supercritical carbon dioxide, and used to catalyse the hydrogenation alkenes under mild conditions. The catalyst can then be recycled without the use of organic solvents. Supercritical carbon dioxide has been used in the removal of stains from textiles, the preparation of biologically active micro-particulate insulin powders and to extract commercially hydrophilic and hydrophobic biochemicals from herbs.

→ Both solvents enable new processes and considerably reduce energy consumption and waste.
→ These green solvents are often easier to separate (from the reactants and products) and recycle

Biofuels and solvents

Transesterification of vegetable oils and fats (triglycerides) into long chain (R) fatty acid esters (biodiesel fuels) and glycerol is carried out in batch reactors or, more recently, in a continuous flow process involving ultrasonics.

$$
\begin{array}{lll}
R_1.CO_2CH_2 & R1.CO_2CH_3 & CH_2OH \\
| & & | \\
R_2.CO_2CH + 3CH_3OH \rightarrow & R_2.CO_2CH_3 + & CHOH \\
| & & | \\
R_3.CO_2CH_2 & R3.CO_2CH_3 & CH_2OH
\end{array}
$$

Pilot-plant scale research in Germany is producing lactic acid from milk whey – an 82 million tonne waste by-product of the milk processing – as an alternative to the current method using molasses from the sugar processing.

Lactic acid is a raw material in the production of ethyl lactate, and other esters, for use as biodegradable solvents in the coatings and paints industry. Ethyl lactate has a high boiling point and low surface tension. It is an effective paint stripper and graffiti remover. Polylactide (PLA), a biodegradable polymer derived from lactic acid, has important biomedical applications (e.g. as a drug encapsulator) because it can be metabolised.

A recent report from Wisconsin University describes research in progress to produce liquid alkanes from the glycerol and water:
Researchers in New Mexico are developing a solar-driven system, christened CR5 (Counter Rotating Ring Receiver Reactor Recuperator) to produce carbon monoxide from carbon dioxide and hydrogen from water. Concentrated sunlight (hv) heats ceramic cobalt-doped ferrite rings to

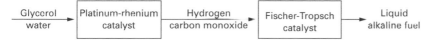

1500 °C releasing oxygen. The reduced ferrite rings are then rotated into an atmosphere of carbon dioxide to produce carbon monoxide and be rotated back into the concentrated sunlight.
A similar process will be used to produce hydrogen from water. The carbon monoxide and hydrogen from the two separate processes can be used to synthesis liquid fuels.

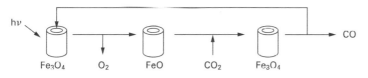

Laboratory on a microchip
Over the past ten years chemists world wide have applied microfluidic technology to miniaturise, integrate and automate many laboratory processes to the so-called lab-on-a-chip (LAC). An LAC may contain a network of microscopic channels (etched in glass, silicon or plastic) along which picolitres of liquids are moved electrically or by pressure under computer control. LAC development was driven initially by genomic research requiring rapid analysis of huge numbers of DNA samples and by drug synthesis research requiring rapid screening hundreds of chemically related substances for biochemical activity. The advantages of LACs (over conventional laboratory work) include

→ very fast using only very small amounts of materials
→ no moving parts - low power consumption and maintenance costs

LACs made from plastic can be disposable to avoid cross-contamination of forensic samples. And they can be automated to provide accurate diagnostic tests in, for example, hospital laboratories.

The jargon

Iodine number (or value) is the mass, in grams, of iodine absorbed by 100 g of a substance (oil, fat, etc.) by addition to double bonds. Typical values for potential biodiesel fuels are 97 for Rapeseed Methyl Ester, 100 for Rapeseed Ethyl Ester, 123 for Soy Ethyl Ester and 133 for Soy Methyl Ester. These values are used to indicate storage instability. European standards set a limit at 120, so the Soy esters would not be used as biodiesel fuels.

Checkpoint 2

If biodiesel and glycerol can be obtained from vegetable oils, state
(a) how this helps reduce atmospheric carbon dioxide levels
(b) prolongs the Earth's oil reserves.

Checkpoint 3

(a) State why lactic acid is a suitable molecule for polymerisation.
(b) Name the class of polymers to which polylactide belongs.

Checkpoint 4

(a) State the name of the lowest member of the alkane homologous series which is liquid at room temperature and atmospheric pressure.
(b) Write a separate balanced equation for each of the two steps in the proposed CR5 process and then combine those two equations to obtain a balanced equation for the overall process.

Action point

Find out what is meant by a picolitre.

Exam practice (6 minutes)   answers: page 193

(a) Write (i) the systematic name of lactic acid, CH3CH(OH)COOH, and (ii) the structural formula of its ethyl ester.
(b) Suggest, in terms of its structure and intermolecular forces, why ethyl lactate has a low volatility and a b.pt. above that of water.

Comprehension question (30 min)

Reproduced, by kind permission of the WJEC

answers: page 193

Read the passage below and then answer the questions (a) to (h) which are based on it.

Pain Killing Drugs

Pain killing or analgesic drugs are in common use both for minor aches and pains and for control of more intense pain in serious medical conditions. The structures of five well-known pain killing drugs are shown below.

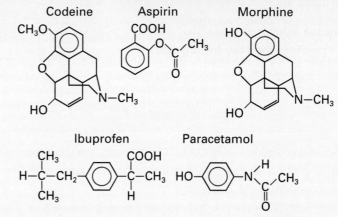

Aspirin and its soluble sodium salt are regarded as the safest of all these drugs, having been in widespread use for so long. Aspirin is made from salicylic acid (2-hydroxybenzenecarboxylic acid), and was first synthesized in 1899. Salicylates, which occur naturally, e.g. in the willow tree, have been used in herbal medicine for a very long time. The salicylates were identified as painkillers in 1876. Aspirin is used for mild pain relief but also reduces inflammation so is used for relief of the symptoms of rheumatoid arthritis in some patients. It is known that, in some cases, aspirin may cause harmful bleeding in the stomach.

Codeine and morphine belong to the opium alkaloids. *Alkaloids* are basic nitrogen-containing naturally occurring compounds, which have pronounced physiological effects on the body. Morphine is the main constituent of opium, a substance derived from the opium poppy. Codeine is a more effective painkiller than aspirin and does not have the addictive and other side effects of morphine. Morphine is used to produce heroin (or *diamorphine*, as it is known in medicine). In heroin the two hydroxy groups of the morphine molecule have been ethanoylated. Heroin is highly addictive and is used mainly for controlling pain in terminally ill patients. Its illegal use and detrimental addictive properties are well documented.

Paracetamol is a readily avaliable painkiller, which, like aspirin, requires no prescription. It is used for mild pain such as headache, toothache and muscle pain. Like all these drugs it is important not to exceed the recommended dosage. Too high a dose of paracetamol can cause harmful physiological effects in the body.

Ibuprofen is the most recent of these drugs. It may be said to be a 'designer drug', having been specifically developed by the Boots Company to alleviate the symptoms of rheumatoid arthritis. Like all drugs that are introduced, it had to undergo extensive testing before approval and before it was available on prescription in 1969. It is now readily available both as non-proprietary ibuprofen or under a variety of proprietary names such as Nurofen.

answers: page 193

(a) State which *one* of the five drugs is considered the safest. Give a reason for your answer.

....

....

(b) (i)   Draw the structure of 2-hydroxybenzenecarboxylic acid.

(ii) State the number of moles of sodium hydroxide which reacts with one mole of aspirin to form the sodium salt.

....

(c) State which of the drugs shown are classified as *alkaloids*.

....

(d) Complete the molecular skeleton given below to show the structure of a molecule of *diamorphine*.

(e) (i)   State the type of isomerism shown by ibuprofen.

....

(ii) Show on the structural formula below the structural feature giving rise to this isomerism.

(f) State which of the five drugs may be classified as a 'designer drug' and give your reasons.

....

....

....

(g) By considering the structure of morphine, explain why
   (i)   in acidic solution the nitrogen atom becomes protonated

....

....

(ii) in alkaline solution the molecule only loses one proton.

....

....

(h) By referring to the drugs listed give
   (i)   *two* advantages to society by the provision of such chemicals

....

....

....

(ii) *one* precaution that must be taken when using thee analgesic drugs.

....

....

Links

See pages 146: functional groups.

Grade booster

Play safe and show *all* the bonds.

Structured exam question

answers: pages 193–4

(a) The molecular formulae of six compounds, A, B, C, D, E and F, are shown in the reaction scheme below. Compound A is a colourless liquid that gives a colourless gas with aqueous sodium carbonate. Compound A can be converted directly to E by refluxing with ethanol and concentrated sulphuric acid.

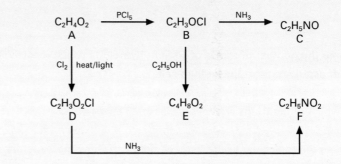

(i) Draw structural formulae for compounds A to F.

A   B   C

D   E   F

Write a balanced equation for the reaction of compound A with

(ii) aqueous sodium carbonate:

....

(iii) ethanol:

....

(iv) Draw a possible structure for a compound formed when F reacts with 2-aminopropanoic acid, $CH_2CH(NH_2)COOH$.

(b) (i) Describe and explain the S_N2 mechanism for the hydrolysis of 1-bromobutane by aqueous sodium hydroxide.

(ii) Write the rate equation for this S_N2 hydrolysis reaction.

....

Links

See pages 146–7: functional groups.

Grade booster

Play safe and show *all* the bonds.

Watch out!

This is one of the very small number of mechanisms required at A-level. Know your mechanisms well if you want the top grade!

Answers
Organic chemistry

How to name compounds

Checkpoints

1 (a) 2-methylbutane   (b) 2,3-dimethylheptane
2 hex-2-ene
   Three isomers
   hex-1-ene
   hex-2-ene
   hex-3-ene
3 1,3,5-trimethylbenzene
   There are four
   1,2,3-trichlorobenzene
   1,2,4-trichlorobenzene
   1,3,4-trichlorobenzene
   1,3,5-trichlorobenzene

Exam practice

(a) They are saturated hydrocarbons having the same molecular formula but different structures and properties.

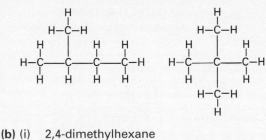

(b) (i) 2,4-dimethylhexane
    (ii) pent-2-ene
    (iii) 4-bromo-2,6-dichloromethylbenzene
(c) (i) $CH_3CH_2CH_2CH_2CH_2CH_2CH_2CH_3$
    (ii)

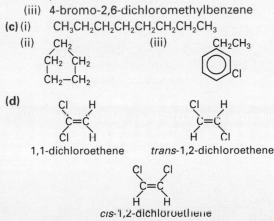

(iii)

(d)

1,1-dichloroethene   *trans*-1,2-dichloroethene

cis-1,2-dichloroethene

Classes of compounds and functional groups

Exam practice

(a) nitrile, alkene, phenyl, chloro.
(b)

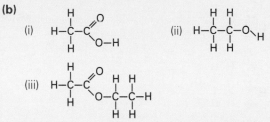

(c) $CH_3COOH + C_2H_5OH \rightleftharpoons CH_3COOC_2H_5 + H_2O$

Types of reactions and reagents

Checkpoints

1 (i)   Electrophile
  (ii) Free radical
  (iii) Nucleophile
  (iv) Nucleophile
2

bromomethane   dibromomethane

tribromomethane   tetrabromomethane

3 (a) (i) An electrophile
      (ii) A nucleophile
  (b) 1,2-dibromoethane.
  The mechanism is electrophilic addition:

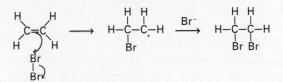

Exam practice

(a) (i) $Cl_2 + h\nu \rightarrow 2Cl\cdot$ initiation
    (ii) $C_2H_6 + Cl\cdot \rightarrow C_2H_5\cdot + HCl$
          $C_2H_5\cdot + Cl_2 \rightarrow C_2H_5Cl + Cl\cdot$ propagation
    (iii) $C_2H_5\cdot + C_2H_5\cdot \rightarrow C_4H_{10}$ termination
    (iv) $C_2H_6 + 2Cl_2 \rightarrow C_2H_4Cl_2 + 2HCl$
(b) The Cl–Cl bond energy is greater than the Br–Br bond energy. Therefore the photon energy, $h\nu$, to break the Cl–Cl bond must be higher than that to break the Br–Br bond. To get a higher photon energy, the frequency (ν) must be greater (h is Planck's constant).
(c) CFCs such as CCl_2F_2 undergo homolysis (homolytic fission) and form chlorine free radicals. The carbon–fluorine bonds do not break because of the very high bond energy of C–F:
$$CCl_2F_2 + h\nu \rightarrow \cdot CClF_2 + Cl\cdot$$
The chlorine free radicals react with ozone molecules
$$O_3 + Cl\cdot \rightarrow O_2 + ClO\cdot$$
The new radical, $ClO\cdot$, enters into a variety of other radical reactions but the net effect is that ozone molecules are destroyed.

The use of CFCs has caused the appearance of holes in the ozone layer in the upper atmosphere. CFCs reach the stratosphere without degradation because they are very stable compounds.

The ozone layer is important because it absorbs harmful UV radiation. Scientists think that depletion of the ozone layer is causing an increase in diseases such as skin cancer by allowing more harmful UV radiation to reach the earth's surface. The EU has now banned the manufacture of CFCs.

Hydrocarbons: alkanes and alkenes

Checkpoint

Each carbon atom in the ethane molecule is surrounded by four bonding pairs of electrons. The arrangement of the bonds around each carbon atom must be tetrahedral.

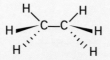

Note that, because there is free rotation about the carbon–carbon bond, the most stable shape is when the end-on view of the molecule is as shown.

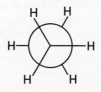

This is called the staggered conformation.

Exam practice

(a) $NaOH + C_4H_9Br \rightarrow C_4H_9OH + NaBr$ is a substitution reaction in which Br is replaced by OH.
The species which attacks the halogenoalkane (C_4H_9Br) is the hydroxide ion, OH^-. The hydroxide ion is a lone pair donor (a Lewis base) or nucleophile. Thus the above reaction is said to be a nucleophilic substitution reaction.

In an addition reaction, a molecule adds on atoms to become a different molecule. The reaction $CH_2=CH_2 + Br_2 \rightarrow CH_2BrCH_2Br$ is an addition reaction. The mechanism involves the heterolysis (heterolytic fission) of the Br–Br bond. This results in the addition of Br^+ to form $CH_2BrCH_2^+$, a carbocation (carbonium ion). Since Br^+ is an electrophile (a lone pair acceptor or Lewis acid) the reaction is called electrophilic addition.

(b) In the absence of UV light.

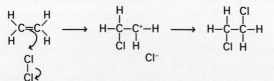

(c) CH_2BrCH_2Br 1,2-dibromoethane
CH_2BrCH_2OH 2-bromoethanol
CH_2ClCH_2Br 1-bromo,2-chloro-ethane

Hydrocarbons: alkenes and arenes

Checkpoints

1

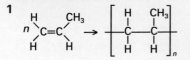

2

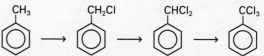

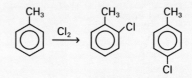

Exam practice

(a) (i) When chlorine is passed into hot methylbenzene, then the side-chain is substituted in a free radical reaction.

(ii) In the absence of heat and light and in the presence of a halogen carrier, a mixture of 2-chloromethylbenzene and 4-chloromethylbenzene is formed.

(b) This reaction is a free radical reaction.
$Cl_2 + h\nu \rightarrow 2Cl\cdot$ initiation
$C_6H_5CH_3 + Cl\cdot \rightarrow C_6H_5CH_2\cdot + HCl$
$C_6H_5CH_2\cdot + Cl_2 \rightarrow C_6H_5CH_2Cl + Cl\cdot$ propagation
Further reaction will occur until all three hydrogen atoms of the side chain have been substituted.

Compounds containing halogens

Checkpoint

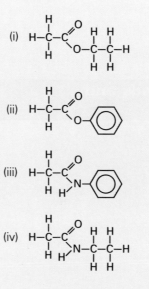

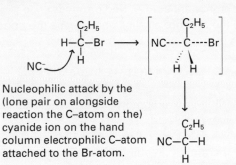

Nucleophilic attack by the (lone pair on alongside reaction the C–atom on the) cyanide ion on the hand column electrophilic C–atom attached to the Br-atom.

+ Br⁻

Compounds containing oxygen: alcohols, phenols, aldehydes and ketones

Checkpoints

1

butan-1-ol (primary) butan-2-ol (secondary)

2-methylpropan-1-ol (primary) 2-methylpropan-2-ol (tertiary)

Grade booster

You are expected to be aware that in both organic and inorganic reactions the product may depend upon the reaction conditions used.

2 $C_6H_5OH + NaOH \rightarrow C_6H_5O^- + Na^+ + H_2O$

Exam practice

(a) Under mild conditions ethanol is oxidized to ethanal.

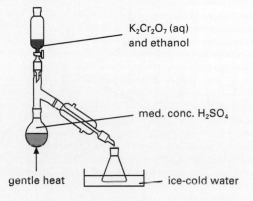

K₂Cr₂O₇ (aq) and ethanol

med. conc. H₂SO₄

gentle heat   ice-cold water

Under harsher conditions, where the reagents are refluxed together, ethanol is oxidized to ethanoic acid.

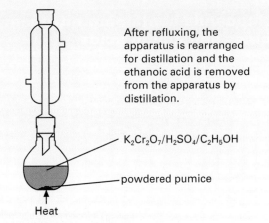

After refluxing, the apparatus is rearranged for distillation and the ethanoic acid is removed from the apparatus by distillation.

$K_2Cr_2O_7/H_2SO_4/C_2H_5OH$

powdered pumice

Heat

(b)(i) Ethene and steam at 70 atm pressure are passed over a catalyst of phosphoric acid on a silica support at 300 °C:
$H_2O(g) + C_2H_4(g) \rightarrow C_2H_5OH(g)$

(ii) When oils or fats are refluxed with aqueous sodium hydroxide, hydrolysis of the ester linkages takes place and results in a mixture of the sodium salts of carboxylic (fatty) acids and glycerol (propan-1,2,3-triol).

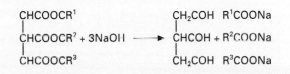

CHCOOCR¹
|
CHCOOCR² + 3NaOH ⟶
|
CHCOOCR³

CH₂COH R¹COONa
|
CHCOH + R²COONa
|
CH₂COH R³COONa

Compounds containing oxygen: aldehydes, ketones and carboxylic acids

Checkpoints

1

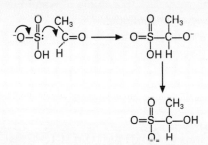

2 $C_2H_5COOH + CH_3OH \rightleftharpoons C_2H_5COOCH_3 + H_2O$

Exam practice

(a)

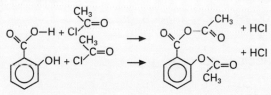

+ HCl

+ HCl

(b) Reflux a mixture of 2-hydroxbenzoic acid with methanol and a small amount of concentrated sulphuric acid to catalyze the reaction and remove water to shift the equilibrium to the right to improve the yield.

Compounds containing nitrogen: amines and amino acids

Checkpoints

1 (i)

H–C–C–C–C–H with H, H, NH₂, H groups

2-aminobutane – primary

(ii)

NH₂ / CH₃ on benzene ring

2-methylphenylamine – primary

(iii)

CH_3 \
N–C_2H_5 \
CH_3

ethyldimethylamine – tertiary

(iv)

CH_3 CH_3 \
N on benzene ring

N:N-dimethylphenylamine – tertiary

2 $CO(NH_2)_2 + 2HNO_2 \rightarrow CO_2 + 3H_2O + 2N_2$

Note: You might expect the product to be $CO(OH)_2$. This is carbonic acid, H_2CO_3, which decomposes to carbon dioxide and water.

3 NH₂ / Br, Br, Br on benzene ring

Exam practice

(a) Phenylamine is only very slightly soluble in water but the salt, $C_6H_5NH_3Cl$, is readily soluble.

(b) The alkali reacts with the phenolic group to give the soluble sodium salt whereas 2-naphthol is insoluble.

(c) Aromatic diazonium ions are stabilized by electrons of the multiple bond between the nitrogen atoms participating in the delocalization of the π-electrons of the ring.

Compounds containing nitrogen: proteins and polyamides

Checkpoints

1 $(CH_3CO)_2O + 2NH_3 \rightarrow CH_3COONH_4 + CH_3CONH_2$

2

$nClOC(CH_2)_8COCl + nH_2N(CH_2)_6NH_2$

↓

$-[NHOC(CH_2)_8CONH(CH_2)_6]_n- + 2n\ HCl$

Exam practice

(a) Tin, concentrated hydrochloric acid and nitrobenzene are refluxed together in an apparatus contained within a fume cupboard. When the reaction is complete, the mixture is allowed to cool. At this stage, the

phenylamine is in the form of phenylammonium hexachlorostannate(IV). Sodium hydroxide is added to make the mixture alkaline and release the phenylamine, which is removed by steam distillation.

Overall

$3Sn + 2C_6H_5NO_2 + 14HCl \rightarrow$
$(C_6H_5NH_3)_2SnCl_6 + 2SnCl_4 + 4H_2O$

The addition of sodium hydroxide releases the phenylamine:

$C_6H_5NH_3^+ + OH^- \rightarrow C_6H_5NH_2 + H_2O$

(b) At temperatures greater than 10 °C, phenol is formed:

$C_6H_5NH_2 + HNO_2 \rightarrow C_6H_5OH + N_2 + H_2O$

Between 0 °C and 10 °C, benzenediazonium chloride is formed:

$C_6H_5NH_2 + 2HCl + NaNO_2 \rightarrow C_6H_5N_2Cl + NaCl + H_2O$

(c)

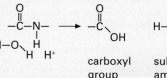

(d) (i) Ammonia is readily evolved:

$RCOONH_4 + NaOH \rightarrow RCOONa + NH_3 + H_2O$

(ii) Primary amines do not react with aqueous sodium hydroxide.

(iii) Ammonia is given off on heating:

$RCONH_2 + NaOH \rightarrow RCOONa + NH_3$

(e) Since nylon is a polyamide, on refluxing with hydrochloric acid the peptide links would be hydrolyzed.

Diagram showing hydrolysis of amide bond producing carboxyl group and substituted ammonium group.

Instrumental techniques: UV, visible and IR spectroscopy

Checkpoints

1

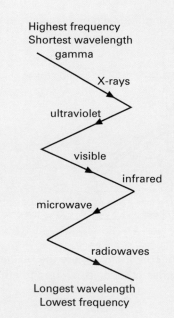

Highest frequency \
Shortest wavelength \
gamma \
X-rays \
ultraviolet \
visible \
infrared \
microwave \
radiowaves \
Longest wavelength \
Lowest frequency

2 (a) The blue end. Bluish light is absorbed so that the pinkish light is transmitted.
(b) You to would be expected to know that chlorophyll is green and therefore to deduce that red and blue light must be absorbed giving two peaks in the red and blue regions. Here is an absorption spectrum of chlorophyll:

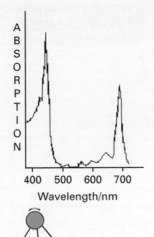

ABSORPTION

400 500 600 700
Wavelength/nm

3

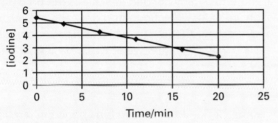

Exam practice

1 The 2nd, 3rd and 5th readings are at 4-min intervals and correspond to the same change in concentration of 0.64×10^{-4} mol dm^{-3}. This suggests concentration varies linearly with time.

[graph of [iodine] against Time/min, showing a straight descending line from about 5.5 at time 0 to about 2.3 at time 20]

The graph of [iodine] against time is a straight line, so the reaction is zero order with respect to iodine.

2 (a) A sample of the contaminated PABA would be scanned in a UV/visible spectrometer and the instrument's computer would compare the result with its database spectrum for pure PABA. A trace contaminant would probably generate unexpected peaks in the sample's spectrum. An IR spectral analysis would be run in the same way.

(b) Samples of exhaust gases (a complex mixture of gases including carbon monoxide, carbon dioxide, nitrogen and oxides of nitrogen together with unburned hydrocarbons and hydrocarbon degradation products) would be withdrawn from within the exhaust system and analyzed
1 by separating the components using gas/liquid chromatography and then
2 analyzing the separate components by infrared spectroscopy.

Instrumental techniques: NMR and mass spectrometry

Checkpoint

The instrument is at low pressure
• to prevent the formation of spurious ions
• so that the ions have an unhindered passage through the apparatus.

Exam practice

1 (a) 31 could be CH_2OH^+
       29 could be $C_2H_5^+$
   (b) 17 OH^+
       15 CH_3^+
   (c) A parent molecule ion containing a ^{13}C isotope
   (d) Infrared
2 Two peaks in the ratio 3 : 1.
There are two types of proton. The ^1H of the —OH group and the three ^1H of the methyl group. The methyl protons provide a peak which is three times the area of that due to the hydroxyl proton.

Analytical techniques: chromatography

Checkpoints

1 (a) thin liquid film bonded to the cellulose surface
   (b) (i) aqueous methanol/acetone (ii) toluene.
2 (a) maximise accuracy of Rf measurement
   (b) samples may be carried to solvent front
   (c) eluent carries spots to same low level as liquid's upward flow is balanced by evaporation
2 Molecules move back and forth between the carrier and stationary phase. Some held more strongly in the stationary phase stay longer there and so move more slowly along the column than the molecules less firmly held.

Exam practice

The components of the sample can be separated by chromatography and just the molecules of each individual component can be injected and separately analysed by the mass spectrometer.

Industrial chemistry: oil refining and petrochemicals

Checkpoints

1 Polyacrilonitrile is an addition polymer:

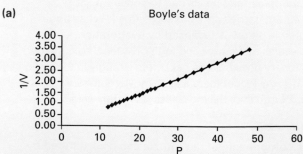

2 When benzene and propene combine, an electrophilic substitution reaction takes place. This is often called a Friedel–Crafts reaction.
Charles Friedel (French) 1832–1899
James C. Crafts (American) 1839–1917

3 A condensation polymer.
Monomers are
1,4-benzenedicarboxylic acid
ethane-1,2-diol

Exam practice

(a)(i) The efficiency of the catalyst falls over a period of time and so it is recycled through a regeneration process to restore its efficiency and to contribute to the economy of the process.

(ii) Hydrocarbon impurities are oxidized.

(b)(i) an alkane (saturated)

(ii) an arene (unsaturated)

How chemistry works 1: data and patterns

Checkpoints

1 E_{zn} = -0.76 V; E_{Ag} = +80 V; E_{cell} = +1.56; so for 12 volts you need 12/1.56 = 8 layers

2 gallium, scanium, geranium

3 Bohr's work on line spectra led to electrons arranged in orbits and Lewis's work on bonding led to electron pairs in oribitals

Exam practice

(a)

Boyle's data

[graph: y-axis $1/V$ from 0.00 to 4.00; x-axis P from 0 to 60, showing a straight line of points rising from approximately (12, 0.8) to (48, 3.3)]

(b) Some doubt about Boyle's published data – lie too precisely on a straight line – little or no sign of scatter expected from experimental error.

How chemistry works 2: principles, theories and applications

Checkpoints

Fuel cells do not store energy;
Fuel and oxidant must be continuously supplied to the cells.

Exam practice

(a) Advantages: greater efficiency, less pollution, potentially a renewable resource.
Disadvantages: storage problems, limited life of expensive catalytic electrodes, adsorbers & adsorbers, toxic chemicals used in their production

(b) ever decreasing size of microchips require finer electrical wire connections, carbon nanotubes could have higher conductivities and lower heat output.

How chemistry works 3: resources and society

Checkpoints

1 Far more thin cans needed to achieve the previous weight so extra collection effort not seen as worthwhile.

2 Sand is denser than water which is denser than oil, so the sand will sink in the water and oil will float on the water's surface.

Exam practice

(a) Assume 22–7 = 15 million to tons buried in landfills. 150 million tons over ten years x 1250 = £187 billion in value buried.

(b) (i) expensive to dig up landfill and separate the non-magnetic cans from the other rubbish

(ii) release methane and other ozone depleting gases into the air (together with other toxins and bacteria).

How chemistry works 4: environmental and global concerns

Checkpoint

Cancel matching species on the LHS and RHS to leave uncancelled species in the overall equation.

$$ClO + ClO \rightarrow Cl_2O_2$$
$$Cl_2O_2 + hv \rightarrow Cl + ClO_2$$
$$ClO_2 \rightarrow Cl + O_2$$
$$2Cl + 2O_3 \rightarrow 2ClO + 2O_2$$
$$2O_3 \rightarrow 3O_2$$

Exam practice

(a) C_3H_8 + $5O_2$ → $3CO_2 + 4H_2O$
44.08   160   132
atom economy CO_2 is 100 x (132/204.08) = 64.7%

(b) Mass of 1 litre is 1000 x 0.585 = 585 g
585 g produces 132 x 585/44.08 = 1752 g CO_2

(c) World Resources Centre calculator gives a lower than theoretical value perhaps to allow for 14% incomplete combustion.

How chemistry works 5: air, earth, fire and water

Checkpoints

1 (a) CO $+$ $3H_2$ $\rightarrow$ $CH_4 + H_2O$

(b) in methane H:C ratio is 4:1 and in alkanes generally it is lower and approaching 2:1

2 (a) H_2O $+$ N_2 $+$ $2\frac{1}{2}O_2$ $\rightarrow$ $2HNO_3$

(b) NH_3 $+$ $2O_2$ $\rightarrow$ HNO_3 $+$ H_2O

Exam practice

A water sample is injected into a clear blue bunsen flame and the resulting colour observed (yellow for sodium and red for calcium). The colour arises from the distinct spectral lines arising from electronic transitions (between ground and excited states) in the gaseous atoms.

How chemistry works 6: the green crystal ball

Checkpoints

1 (a) propane-1,2,3-triol

(b) explosives

(c) the hydrogen adds across the double bond and converts the unsaturated oil into a saturated fat with a higher softening point.

2 (a) the carbon dioxide produced when the fuel burns is 'recycled' by being taken up (together with water) by the plant during phtosynthesis of its carbohydrates.

(b) the green fuel (from renewable sources) is used instead of the non-rewable petroleumt.

3 (a) the molecule has two functional group (–CO$_2$H and –OH) that can interact

(b) polyester

4 (a) pentane

(b) $Fe_3O_4 \rightarrow 3FeO + \frac{1}{2}O_2$; $3FeO + CO_2 \rightarrow Fe_3O_4 + CO$

$CO_2 \rightarrow CO + \frac{1}{2}O_2$

Exam practice

(a) (i) 2-hydroxypropanioc acid

(ii)

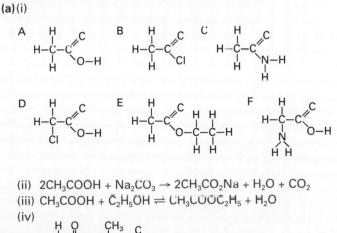

(h) r m m. 118 compared to 18 for water so van der Waals forces would be higher and H-bonding also possible although lower than that between water molecules.

Comprehension question

(a) Aspirin has been in use for a long time.

(b) (i)

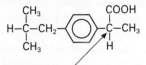

(ii) One

(c) Codeine and morphine

(d)

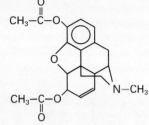

(e) (i) Optical isomerism

(ii) Asymmetric carbon atom clearly indicated.

(f) Ibuprofen was specially produced to treat rheumatoid arthritis.

(g) (i) The nitrogen atom has a lone pair of electrons to combine with H$^+$.

(ii) Only the phenolic —OH group is acidic; the other —OH group is alcoholic/is not phenolic group.

(h) (i) This is a fairly open ended question so we would give you marks for any reasonable comments referring to advantages to society: e.g. control of severe pain in serious medical conditions, to make patient more comfortable, to enable minor discomfort not to affect the quality of daily life, to prevent loss of working day by alleviating minor aches and pains. Use of non-prescription drugs for self-medication saving GP's time, etc.

(ii) Not to exceed the stated dose could be linked to addictive nature of some drugs.

Structured exam question

(a) (i)

(ii) $2CH_3COOH + Na_2CO_3 \rightarrow 2CH_3CO_2Na + H_2O + CO_2$

(iii) $CH_3COOH + C_2H_5OH \rightleftharpoons CH_3COOC_2H_5 + H_2O$

(iv)

Grade booster

This is one of two possible dipeptides. Draw the other. Either one will get you two marks.

(b)(i) Substitution nucleophilic bimolecular:

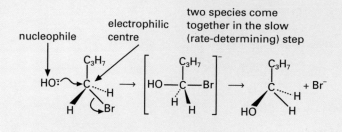

(ii) Rate equation is: rate = $k\,[OH^-][RBr]$.

Resources

This section is intended to help you develop your study skills for exam success. You will benefit if you try to develop skills from the beginning of your course. Modern A-level exams are not just tests of your recall of textbooks and your notes. Examiners who set and mark the papers are guided by assessment objectives that include skills as well as knowledge. You will be given advice on revising and answering questions. Remember to practise the skills.

Exam board sepcifications

In order to organize your notes and revision you will need a copy of your exam board's syllabus specification. You can obtain a copy by writing to the board or by downloading the syllabus from the board's web site

AQA (Assessment and Qualifications Alliance)
Publications Department, Stag Hill House, Guildford, Surrey GU2 5XI –
www.aqa.org.uk

CCEA (Northern Ireland Council for Curriculum, Examinations and Assessment)
Clarendon Dock, 29 Clarendon Road, Belfast BT1 3BG – www.ccea.org.uk

EDEXCEL
Stewart House, 32 Russell Square, London WC1B 5DN – www.edexcel.org.uk

OCR (Oxford, Cambridge and Royal Society of Arts)
1 Hills Road, Cambridge CB2 1GG – www.ocr.org.uk

WJEC (Welsh Joint Education Committee)
245 Western Avenue, Cardiff CF5 2YX – www.wjec.co.uk

Topic checklist

Tick each of the boxes below when you are satisfied that you have mastered the topic

	Edexcel		AQA		OCR		WJEC		CCEA	
	AS	A2	AS	A2	AS	A2	AS	A2	AS	A2
Studying and learning	O	●	O	●	O	●	O	●	O	●
Searching the web and Looking for books	O	●	O	●	O	●	O	●	O	●
Nobel Prize Winners	O	●	O	●	O	●	O	●	O	●
Taking examinations	O	●	O	●	O	●	O	●	O	●
Synoptic assessment 1–4: Your examination papers and assessment objectives	O	●	O	●	O	●	O	●	O	●
Data handling 1 & 2: calculations, problem solving and exam questions	O	●	O	●	O	●	O	●	O	●

Studying and learning

How good are your study skills? Answer these questions.
1 Do you enjoy studying chemistry?   yes ☐ no ☐
2 Do you have good study habits?   yes ☐ no ☐
3 Do you follow a programme of spaced revision?
                                     yes ☐ no ☐
4 Do you regularly tackle chemistry questions?
                                     yes ☐ no ☐
5 Do you take practice papers under exam conditions?   yes ☐ no ☐

What is study?

Physical activities (writing, drawing, calculating, computing, word-processing, etc.) produce something you can see: e.g. a neat set of notes, your account of an experiment, your solution to a problem. This tangible evidence of your hard work can pile up, give you a sense of satisfaction and enhance your self-esteem. But there are other, perhaps more important, activities.

Mental activities (reading, listening, recognizing, recalling, reflecting, remembering, knowing, thinking, etc.) produce no tangible evidence of your hard work until you demonstrate to yourself and others what you have understood and learnt as a result of your studying. You do this when you get high marks in tests and exams. But you don't take a test or exam every day so there is a danger of spending too much time on physical activities.

Smart study

You will have more written notes at the end than at the start of your course. Don't aim to acquire and retain a huge pile of notes. Keep their volume to a minimum by revising and condensing them regularly. Make mind maps.

➔ Aim to understand the facts, patterns, principles and theories of chemistry.
➔ Combine mental and physical activities by tackling questions and problems.

We are usually better at subjects we like than at subjects we dislike. The more we like a subject the easier we find it and the more time we spend studying it. A habit is something that doesn't need great effort or will-power. Good students never seem to have to think about studying because they habitually use good study techniques. To build good study habits, top students

➔ do private study in the same place at the same times each day
➔ warm up at the start of each study session with a simple task
➔ work up gradually from easier to more difficult parts of a topic
➔ stop each study session while they are still enjoying success
➔ clear their table and leave a simple task to start their next session

Here is the psychology behind this five-point plan. As soon as you sit at your table you have pleasant memories of your previous study sessions. There on your table is something simple for you to do and there

The jargon

URI (or URL) = unique resource identifier (or locator)
http = HyperText Transfer Protocol
www = world wide web – a dynamic information-sharing system using the huge interconnected networks of computers (the *internet*) situated round the world. It is not a static system and websites are changing all the time.

The jargon

Mind mapping is a technique devised by Tony Buzan for improving the way you take notes. A mind map shows the structure and links between points of a topic in a format that you can review quickly and remember easily.

Action point

Search on the world wide web for *mind mapping* and learn how to make notes in the form of mind maps.

Check the net

You can obtain a wealth of information on the internet. Here are a few places to start looking:
www.chemweb.com
www.chemsoc.org
www.acdlabs.com/downloads

Watch out!

One hour of effective studying that you enjoy is worth ten hours of ineffective studying that you dislike. Aim to keep yourself mentally and physically fit.

is nothing to distract you. So as a matter of habit you immediately find yourself doing something useful. In no time at all, without any conscious act of will, you are automatically engrossed in your studies.

→ Never study for so long that you stop because you are tired, fed up and no longer enjoying your work.

Various investigations and research have shown that top students

→ revise material as soon as possible
→ use frequent and short periods of revision
→ carefully space their periods of revision

Tony Buzan, an expert on study techniques and author of the book *Make the Most of Your Mind*, recommends a pattern of spaced revision to follow a one-hour study session.

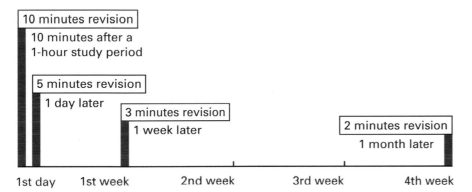

Really good students know the benefit of preparing for examinations by regularly tackling questions and practice papers, then correcting their answers very carefully. In this way they get to know the language the examiners use and have a much clearer idea of what the examiners want for a good answer.

Learning styles

We gather information by seeing, visualizing, listening, speaking, moving and doing things. How we process the information depends on our learning styles which may involve:

→ working with concrete examples or dealing with abstract concepts
→ actively experimenting or reflecting on observations
→ arriving at an understanding by a logical sequence of small steps or by starting with a general impression and filling in the details

Top students reflect on their experiences (and those of others), form opinions and reach conclusions which they can try out. These trials give them new experiences to reflect upon and keeps the learning cycle turning.

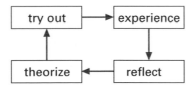

Top students know there is always more to learn and they use the learning styles most appropriate to each task.

Grade booster

Study and revise actively by processing the information. In your own textbook, underline or highlight key points and write cryptic notes in the margins. Convert headings into key questions. Condense text into a diagram, graph, picture or a mind map.

"We only think when we are confronted with a problem."

John Dewey

The jargon

Visual, auditory and *kinaesthetic* are the three modes for gathering information.

Action point

Search on the world wide web for *learning styles* and visit the study skills section of the BBC web site at www.bbc.co.uk/education/asguru/

Searching the web

The world wide web is a way of accessing information over the internet. You should learn to browse and search the web efficiently to avoid aimless surfing and to avoid being overloaded with information.

Techniques for accessing the web

Browsing and Searching

A web page usually contains *links* to other pages on the same site or on other web sites. You *browse* or *surf* the web when you click these links and follow the trail from one page to the next. Browsing or surfing is appropriate to
→ check out what is available in a specific subject area
→ get ideas for conducting a more precise search
→ start a new search if a previous search gave unsatisfactory results

Microsoft Internet Explorer and Netscape Navigator are two popular free multimedia (graphics, sound and video) browsers. You can record the URLs of web pages you may wish to revisit and store web pages for later viewing offline.

You *search* the web for a keyword or phrase using a *search engine*. Searching is appropriate when
→ you need to locate a specific item you know is on the web
→ you need specific information you can define by precise search criteria

AltaVista and **Google** are popular search engines usually yielding an enormous number of irrelevant results by treating all sites equally, including personal and recreational websites. **ChemFinder** (http://chemfinder.cambridgesoft.com) is a more specific engine to find chemicals by name, properties or structure.

Yahoo is a well-known general and specialized *Internet directory* for browsing by category or searching by keyword the records describing websites rather than the contents of the sites themselves. So you can use broad search terms and reduce the risk of being swamped by the results.

PSIgate and **Chemsoc** are *information gateways* to subject-specific resources of quality and relevance in the physical sciences and chemistry. You might find fewer resources with a gateway compared to a search engine or directory but the information should be highly relevant and reliable.

Words of warning

→ Evaluate a resource before you use it – check where, when and why the information was published and by whom.
→ Don't neglect your textbooks and other established sources of information.

This Study Guide will help you achieve the top A-level grade. We advise you to look also at publications some Awarding Authorities produce for their specific examination and include in their specifications as suggested reading. You will find on the page 199 details of these publications and other books to help you with your studies. You should find many more in book shops, at your local library and online at Amazon (http://www.amazon.co.uk).

The jargon

Web page – document on the world wide web (www) identified by a unique Uniform Resource Locator (URL).
Web browser – software application used to find and view web pages.
Link or *hyperlink* – item on an electronic page connecting to another item somewhere else (on the same page, on another page in the same document or an entirely different one).
Search engine – a system or class of programs for locating a word or phrase in electronic documents on the web.
Web portal or *gateway* – site providing descriptions of and links to a collection of selected resources on the web.

Grade booster

You can practise your internet information skills with a the free tutorial at Internet Chemist (http://www.vts.rdn.ac.uk/tutorial/chemistry) and improve your searching techniques with the Danny Sullivan's Search Engine Math page (http://www.searchenginewatch.com/facts/math.html).

Watch out!

Spelling mistakes (ethene instead of ethane) can be a disaster. Sometimes even the wrong case (Ethane instead of ethane) might spoil your search.

Action point

Visit Yahoo (at http://uk.yahoo.com) and follow the web directory links Science → Chemistry → Web Directories. From this page follow the links to explore ChemFinder and Sheffield Chemdex.

Action point

On PSIgate (http://www.psigate.ac.uk) go to the chemistry gateway and in the browse chemistry resources take the link general chemistry → Adrian Dingle → links (The best of Chemistry on the Web) → Chemguide to browse Jim Clark's website.

Watch out!

If you can't bookmark or add good sites to your favourites, cut and paste their URLs into a file but don't forget that information and sites can change, move or disappear.

Looking for books

Awarding Authority publications

Authority	Title	Author	Publisher
Edexcel	Units 1,2, 4 and 5	Beavon and Chapman	Nelson
	Nuffield Advanced Chemistry	Vokins	Longman
	Nuffield Chemistry: Book of Data		Longman
OCR	Chemistry 1	Ratcliff, Eccles et al	Cambridge U. P.
	Revise AS Chemistry for OCR	Eccles	Heinemann Educational
	Introduction to Advanced Chemistry	Earl and Wilford	John Murray
	AS Chemistry	Hunt	Hodder and Stoughton
	Revise AS Chemistry	Ritchie	Letts
OCR Chemistry Options support material	Methods of Analysis Biochemistry Environmental Chemistry Gases, liquids and solids Transition elements	McCarthy Harwood Winfield Matthews Acaster	Cambridge University Press
	Salters' Chemistry support books		Heinemann Educational Secondary Division
WJEC	Contact WJEC website		

Popular texts

Title	Author	Publisher
Chemistry in Action	Freemantle	Thomson Learning
Chemistry in Context	Hill, Holman	Nelson
A Level Study Aids- Chemistry	McMonagle	John Murray
A level Chemistry	Ramsden	Stanley Thornes
Modern Inorganic Chemistry	Liptrot	Collins
Modern Organic Chemistry	Norman and Waddington	Collins
Modern Physical Chemistry	Liptrot et al	Collins
Chemical Principles	Zumdahl	Houghton Miflin Co

Grade booster

Students gaining the highest marks often reveal by their answers that have read other textbooks besides those issued by their school or college.

Action point

Get a set of textbooks of your own so you can highlight key points and make notes in the margins (see the section on studying and learning on page 196). Search the Amazon website for reviews, sample pages and second-hand bargains before you buy. Each of the Awarding Authorities produce student support material dedicated to that specification.

Watch out!

The list is not exhaustive and only you can decide the usefulness of these publications. The WJEC Revision Aids and Background Reader are also available in Welsh.

Grade booster

Steven S. Zumdahl, Professor of Chemistry at the University of Illinois, has received awards for excellence in teaching and for his contributions to chemical education. *Chemical Principles* is one of many American chemistry textbooks he has written.

Action point

Ask your teachers for their recommended book list and for a copy of your course specification – it will be invaluable for revision. Visit the QCA and your Awarding Authority websites (see Synoptic Asessment on page 202).

Nobel Prize winners
1901–2007

The last hundred years have seen great advances in chemistry which have affected our everyday lives.

The contributions of chemistry to society

We live in a chemical age. The advances in chemistry can be summarized in the list of the Nobel Prize Winners for Chemistry through the twentieth and twenty first centuries.

Where two or more names appear in the list for a particular year, then the prize was shared.

Action point

Go through the list of winners and mark any whose names appear in the rest of the book. Surf the web for their biographies and details of their work. eg. go to http://www.almaz.com/nobel/chemistry/

Chemistry Nobel Prize Winners 1901–2007

Year	Prize winners	Subject
2007	Gerhard Ertl	chemical processes on solid surface
2006	Roger D Kornberg	molecular basis of eukaryotic transcription
2005		
2004	Aaron Ciechanover, Avram Hershko, Irwin Rose	discover of ubiquitin-mediated protein degradation
2003	Peter Agre, Roderick MacKinnon	discoveries concerning channels in cell membranes
2002	John B. Fenn, Koichi Tanaka, Kurt Wuthrich	analyses of biological macromolecules NMR spectroscopy of biological macromolecules
2001	William S Knowles, Ryoji Noyori, K. Barry Sharpless	chirally catalyzed hydrogenation chirally catalyzed oxidation reactions
2000	Alan J. Heeger, Alan G. MacDiarmid, Hideki Shirakawa	conductive polymers
1999	Ahmed Zewail	femtosecond spectroscopy
1998	Walter Kohn, John A. Pople	theoretical chemistry
1997	Paul D. Boyer, John E. Walker, Jens C. Skou	synthesis ATP discovery of an ion-transporting enzyme
1996	Robert F. Curl, Jr., Sir Harold W. Kroto, Richard E. Smalley	discovery of the fullerenes
1995	Paul Krutzen, Mario Molina, F. Sherwood Rowland	ozone depletion in the atmosphere
1994	George A. Olah	carbocation chemistry
1993	Kary B. Mullis, Michael Smith	invention of the polymerase chain reaction contributions to protein studies
1992	Rudolph A. Marcus	electron-transfer reactions
1991	Richard R. Ernst	NMR spectroscopy
1990	Elias James Corey	organic synthesis
1989	Sidney Altman, Thomas R. Cech	catalytic properties of RNA
1988	Johann Deisenhofer, Robert, Huber, Hartmut Michel	three-dimensional structure of a photosynthetic reaction centre
1987	Donald J. Cram, Jean-Marie Lehn, Charles J. Pedersen	structure-specific interactions of high selectivity
1986	Dudley R. Herschbach, Yuan T. Lee, John C. Polanyi	dynamics of chemical elementary processes
1985	Herbert A. Hauptman, Jerome Karle	methods for the determination of crystal structures
1984	Robert B. Merrifield	chemical synthesis on a solid matrix
1983	Henry Taube	electron transfer in metal complexes
1982	Sir Aaron Klug	crystallographic electron microscopy
1981	Kenichi Fukui, Ronald Hoffmann	theories in theoretical chemistry developed independently
1980	Paul Berg	recombinant DNA
	Walter Gilbert, Frederick Sanger	base sequences in nucleic acids
1979	Herbert C. Brown, Georg Wittig	boron and phosphorus chemistry in organic synthesis
1978	Peter D. Mitchell	chemiosmotic theory
1977	Ilya Prigonine	stereochemistry non-equilibrium thermodynamics
1976	William N. Liscombe	the boranes
1975	Sir John Warcup Cornforth, Vladimir Prelog	stereochemistry of enzyme-catalyzed reactions stereochemistry of organic molecules
1974	Paul J. Flory	macromolecules
1973	Ernst Otto Fischer, Sir Geoffrey Wilkinson	organometallic sandwich compounds
1972	Christian B. Anfinsen	amino acid sequencing
	Standford Moore, William H. Sein	catalytic activity of the active centre of the ribonuclease molecule
1971	Gerhard Herzberg	structure and geometry of free radicals
1970	Luis F. Leloir	sugar nucleotides
1969	Sir Derek H. R. Barton, Odd Hassel	concept of conformation and its application
1968	Lars Onsager	thermodynamics
1967	Manfred Eigen, Ronald George	fast chemical reactions

	Norrish, Lord George Porter	
1966	Robert S. Mulliken	molecular orbital chemistry
1965	Robert Burns Woodward	organic synthesis
1964	Dorothy C. Hodgkin	X-ray techniques of biological compounds
1963	Karl Ziegler, Giulio Natta	polymer catalysts
Year	*Prize winners*	*Subject*
1962	Max Ferdinand Perutz, Sir John C. Kendrew	study of globular proteins
1961	Melvin Calvin	CO_2 assimilation in plants
1960	Willard Frank Libby	carbon-14 dating
1959	Jaroslav Heyrovsky	polarographic analysis
1958	Frederick Sanger	work on proteins and especially insulin
1957	Lord Alexander Todd	nucleotides and nucleotide co-enzymes
1956	Sir Cyril Hinshelwood, Nikolay Semenov	reaction mechanisms
1955	Vincent du Vigneaud	first synthesis of a polypeptide hormone
1954	Linus C. Pauling	nature of the chemical bond
1953	Hermann Staudinger	macromolecular chemistry
1952	Archer J. P. Martin, Richard L. M. Synge	discovery of partition chromatography
1951	Edwin McMillan, Glenn T. Seaborg	chemistry of the transuranium elements
1950	Otto Paul Diels, Kurt Alder	diene synthesis – the Diels-Alder reaction
1949	William Francis Glauque	low temperature thermodynamics
1948	Arne Wilhelm Kaurin Tiselius	electrophoresis and serum proteins
1947	Sir Robert Robinson	alkaloid chemistry
1946	James Batcheller Sumner, John H. Northrop, Wendell M. Stanley	crystallizing enzymes pure enzymes and virus proteins
1945	Artturi I. Virtanen	agricultural and nutrition chemistry
1944	Otto Hahn	discovery of nuclear fission
1943	George de Hevesy	isotopes as tracers
1940–42	**No prize awarded**	
1939	Adolf F. J. Butenandt	work on sex hormones
	Leopold Ruzicka	polymethylenes and higher terpenes
1938	Richard Kuhn	carotenoids and vitamins
1937	Sir Walter N. Haworth, Paul Karrer	vitamins A and B2
1936	Petrux J. W. Debye	dipole moments and X-ray diffraction
1935	Frederick Joliot, Irene Joliot-Curie	synthesis of new radioactive elements
1934	Harold C. Urey	discovery of heavy hydrogen
1933	**No prize awarded**	
1932	Irving Langmuir	surface chemistry
1931	Carl Bosch Friedrich Bergius	chemical high pressure methods
1930	Hans Fischer	haemin and chlorophyll
1929	Sir Arthur Harden, Hans von Euler-Chelpin	fermentation processes and enzymes
1928	Adolf Otto Windaus	sterols and vitamins
1927	Heinrich Otto Wieland	bile acids
1926	Theodore Svedberg	disperse systems
1925	Richard A. Zsigmondy	colloid chemistry
1924	**No prize awarded**	
1923	Fritz Pregl	organic microanalysis
1922	Francis W. Aston	mass spectrometry and isotopes
1921	Frederick Soddy	radioactivity
1920	Walther H. Nernst	thermodynamics
1919	**No prize awarded**	
1918	Fritz Haber	synthesis of ammonia
1916–17	**No prize awarded**	
1915	Richard Martin Willstätter	chlorophyll and plant pigments
1914	Theodore William Richards	accurate determinations of atomic weights
1913	Alfred Werner	work in bonding in inorganic chemistry
1912	Victor Grignard,	organometallic chemistry – Grignard reagents
	Paul Sabatier	catalytic hydrogenation
1911	Marie Curie	discovery of radium and polonium
1910	Otto Wallach	chemical industry and alicyclic compounds
1909	Wilhelm Ostwald	catalysis/equilibria/kinetics
1908	Lord Ernest Rutherford	radioactivity
1907	Eduard Buchner	biochemistry/cell-free fermentation
1906	Henri Moissan	fluorine chemistry
1905	Johann von Baeyer	organic chemistry
1904	Sir William Ramsay	discovery of the inert gases
1903	Svante Arrhenius	electrolytic theory
1902	Herman Emil Fischer	sugar and purine syntheses
1901	Jacobus Van't Hoff	chemical dynamics and osmotic pressure

Taking examinations

You need to understand the examiners' language and to know what they want. A good exam technique can make the difference between one grade and another.

Grade booster

Quality of written communication must be assessed at AS and A2-level by all exam boards. So examiners must check that you write legibly, spell correctly, punctuate accurately, construct grammatically sound sentences, use appropriate jargon and assemble coherent logically reasoned arguments to clarify your meaning.

Watch out!

Write legibly. If we cannot read it we cannot mark it. Write what you mean and mean what you write. Get help from *Write Right!* by Jan Venolia (ISBN 0-946537-57-7) or *How to Write and Speak Better* by Dr J. E. Kahn *et al.* (ISBN 0-276-42030-6).

Action point

Make and use a set of small cards to help memorize the basic chemical facts, definitions, patterns, etc. Here are three examples:

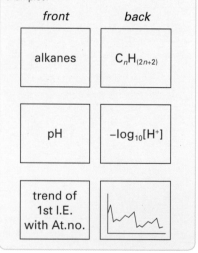

Examiner's secrets

Suggest usually indicates an open-ended question with more than one correct answer that you must find by applying your knowledge and understanding of chemistry.

What are examiners looking for?

Questions are set for you to show your knowledge and understanding of fundamental facts, patterns, principles and theories of chemistry and to demonstrate that you can

→ write and balance equations
→ do calculations (pH, ΔH, K_c, E, etc.)
→ predict the feasibility of reactions
→ deduce organic structures

In the synoptic assessment you can draw upon your knowledge and understanding of all the individual modules to reveal your grasp of AS and A2 chemistry as a whole. When you answer the questions you will inevitably display your *quality of written communication*.

To help you understand the questions, here is a list of the more common words and phrases examiners use.

Concise answers with the bare minimum of detail

Classify each of the following oxides as acidic, basic or amphoteric.
Define the term molar first ionization energy.
Give the oxidation number of uranium in the compound.
Indicate the conditions needed to increase the equilibrium yield.
Name the mechanism for ammonia reacting with bromoethane.
State Hess's law.
What is meant by a buffer solution?
Write a balanced equation for the complete combustion of ethanol.

Concise answers with essential but rather more detail

Calculate the activation energy from the data provided.
Comment on the difference in physical properties of CO_2 and SiO_2.
Deduce the structure of the compound from the information provided.
Draw a labelled Born–Haber cycle for the formation of calcium oxide.
Identify the compounds X, Y and Z in the following observations.
Outline a laboratory method of measuring a named enthalpy change.
Show how you would detect the presence of sodium in a compound.
Sketch the unit cell of a body-centred cubic structure.

Longer answers reasoned with facts and principles

Explain why ammonia is basic and forms complexes with cations.
Explain what is meant by fractional distillation.
State and explain the effect of temperature upon reaction rates.
Suggest how to distinguish 1-bromobutane from 2-bromobutane.

You could find all these types of question in your synoptic papers.

The examination

Preparing for and taking exams

→ Check the regulations to see what you may take into the exam room.

For some papers you may be given a data sheet as well as the periodic table. You may have to use the *Nuffield Advanced Science Book of Data*. This is particularly helpful when you are doing calculations in physical chemistry.

→ In an objective test (OT) paper follow the instructions for correcting a mistake you make.

In an OT paper if you omit a question or you mark more than one choice you will score zero for that particular question. You will NOT score –1 for a wrong choice.

→ In a written paper do not use an 'erasable pen' or white correcting fluid.

If you think you have made a mistake, cross it out neatly with one ruled line. Examiners look at your 'mistakes' and may sometimes be allowed to award you marks for what you have crossed out.

→ Attempt all the compulsory questions.

If you are running out of time, you may gain more marks by answering two questions incompletely than one question completely. You will score zero for any question you do not answer.

→ Tackle at least one set of past or specimen papers.

Use the number of marks and the time allowed for each paper to calculate the *mark rate*: it is often about 1 mark per minute.

→ Take off your wristwatch, put it in front of you where you can see it clearly and get into the habit of checking to keep to time.

If you are running out of time towards the end of an exam, abandon sentences and write your answers in note form even if you run the risk of losing marks for quality of written communication.

The results

We all know how much your results could affect your life. Most students get the result they deserve. If you find that your result is unexpected, you can appeal to the exam board and ask for

→ a clerical check (to see that your marks have been added up and entered into the computer correctly)
→ a re-mark (usually by the Chief Examiner or a senior examiner)
→ a re-mark with a written report

A final word

Self-motivation is the key and is as important as ability. This book, your teachers and fellow students can help but your success depends upon you. Set yourself achievable goals. Develop good study habits. Practise your exam techniques. Work hard. Play hard. And above all, enjoy your chemistry.

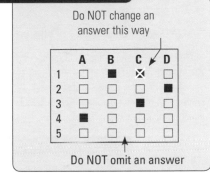

Synoptic Assessment

Greek *syn* - together, *opsis* - seen. Synoptic chart is a weather map showing details of temperature, pressure, amount of cloud, etc., so they can be 'seen together'.

Check the net

Visit the Qualifications and Curriculum Authority website at www.qca.org.uk
Awarding Authority Sites:
www.ccea.org.uk
www.edexcel.org.uk
www.neab.ac.uk
www.ocr.org.uk
www.wjec.co.uk

The jargon

Subject criteria are the instructions to the Awarding Authorities and examiners responsible for the papers you take. There is no overall number of marks given for synoptic assessment

Examiner's secrets

Don't be put off by a question containing material you have never encountered before. The examiners are not allowed to set questions not covered by the specification. You may find the material is unfamiliar but you will be able to answer the question by applying your knowledge and understanding of the chemistry you learnt in your course.

> *"A journey of a thousand miles must begin with a single step."*
> CHINESE PROVERB

Watch out!

Calculators must be silent, cordless and unable to store and display text or graphics. The memory of a programmable calculator must be cleared. Calculator instruction manuals are not allowed into the exam room

The risk with specifications and exams based on modules is that you might learn each module and pass each modular exam without connecting the parts and seeing the subject as a whole. The synoptic assessment was introduced for you to show examiners your holistic understanding of the subject.

A-level specifications and synoptic assessment

Subject criteria

The Qualifications and Curriculum Authority (QCA) has published (on the web) subject criteria for A-level Chemistry and clarified the intention of synoptic assessments to "encourage the development of a holistic understanding of the subject" and support learning. So you should be prepared to use, in contexts which may be new to you, skills and ideas that permeate chemistry, e.g.: writing chemical equations, quantitative work, relating empirical data to knowledge and understanding. You should also be prepared to make and use connections within and between different areas of chemistry, for example, by:

→ applying your knowledge and understanding of principles and concepts of more than one area to a particular situation or context;
→ using your knowledge and understanding of principles and concepts in planning your experimental work and analysing and evaluating your data.

AQA - Synoptic Assessment and Stretch and Challenge

Synoptic assessment in Chemistry is assessed in all the A2 units through both the written papers (Unit 4 and Unit 5) and through the written part of Practical and Investigative Skills in A2 Chemistry (Unit 6). A variety of stems in questions will avoid a formulaic approach by using words such as: analyse, evaluate, compare, discuss. Connections between different content areas will be used where possible and appropriate. Some extended writing will be required and a range of question types (not just short answer/structured questions) will be used to assess different skills.

EDEXCEL - Synopticity

Synopticity requires students to connect knowledge, understanding and skills acquired in different parts of the Advanced Level GCE course. For example, problems related to organic chemistry in Unit 5 require all of the knowledge and understanding the students have developed throughout the other AS and A2 topics. In the examinations at A2 there will be three sections. Sections B and C will contain extended answer questions where students can demonstrate the scientific knowledge they have developed over the whole GCE in Chemistry. These sections will address the synoptic assessment of the GCE in Chemistry. In addition there will be specific questions in Section C of the examinations for Units 4 and 5 which will address synopticity.

OCR - Synoptic Assessment (A Level GCE)

Synoptic assessment tests the candidates' understanding of the connections between different elements of the subject. Synoptic assessment involves the explicit drawing together of knowledge, understanding and skills learned in different parts of the Advanced GCE course. All A2 units, F324–F326, contain some synoptic assessment.

WJEC - Synoptic Assessment (A Level GCE)

Spectroscopy and Organic Chemistry

Sections A (3 structured questions, one based on a short passage) and B (2 extended-answer questions), no choice. Synoptic assessment included.

Physical and Inorganic Chemistry

Sections A (3 structured questions, one based on a short passage) and B (2 extended-answer questions), no choice. Synoptic assessment included. The practical work in CH6 is considered to be synoptic due to the bringing together of knowledge from different areas of the subject and the use of a variety of acquired skills. Synoptic questions in particular may incorporate concepts and ideas which are designed to be more challenging for candidates. The synoptic assessment counts for 20% of your overall A level result.

CCEA - Synoptic Assessment (A Level GCE)

The A2 assessment units include an element of synoptic assessment. In GCE Chemistry, synoptic assessment requires students to make and use connections within and between different areas of the subject at AS and A2, for example by:

→ applying knowledge and understanding of more than one area to a particular situation or context;
→ using knowledge and understanding of principles and concepts in planning experimental and investigative work and in the analysis and evaluation of data; and
→ bringing together scientific knowledge and understanding from different areas of the subject and applying them.

The A2 assessment units will include opportunities for stretch and challenge.

What should you do?

→ Expect to find unfamiliar topics in a question
→ Try to apply what you know and understand to new situations
→ Prepare yourself by practising synoptic chemistry questions
→ Get help from your teachers and your fellow students

A good exam technique can make the difference between one grade and another. For the best grade you must

→ keep on top of your work throughout your A2 year
→ revise thoroughly the facts, patterns, principles and theories you learnt in your AS course.

Do you know how your Board will comply with the QCA specifications for the synoptic assessment of your course? Check carefully your own hard copy of the syllabus or go online to your Board's website.

Don't forget!

You answer objective test questions on a grid using a pencil. There is only one correct answer for each question. A machine marks your paper, so make sure you know the proper way to choose your answer and to alter your answer if you change your mind.

"If you don't know where you are going, you might wind up someplace else."

LAWRENCE PETER "YOGI" BERRA

Examiner's secrets

Do not spend too long on one question and not enough time on another. Time lost can rarely be recovered. Misjudging the time is one of the common mistakes you must avoid if you want the top grade. Never use white correcting fluid because your script would not be valid for an appeal.

Your examination papers and assessment objectives

You should know the form of your examination papers and your assessment objectives.

Assessment Objective Weightings

AQA

Unit 1 4 - 5 short answer questions plus 1 longer structured question. 1¼ hours 33⅓ % of the total AS marks 16⅔ % of the total A-level marks

Unit 2 6 - 8 short answer questions plus 2 longer structured questions. 1 ¾ hours 46 ⅔ % of the total AS marks 23 ⅓ % of the total A-level marks

Unit 3 AS Centre-Assessed Unit 20% of total AS marks 10% of total A-level marks

Unit 4 6 - 8 short answer questions plus 2 structured questions. Some questions will have synoptic elements. 1¾ hours 20% of the total A-level marks

Unit 5 5 - 7 short answer questions plus 2- 3 longer structured questions. Some questions will have synoptic elements. 1¾ hours 20% of the total A-level marks

Unit 6 A2 Centre-Assessed Unit 10% of total A-level marks

CCEA

Unit AS1 1 hour 30 minutes Section A, containing 10 multiple choice questions and Section B containing a number of structured questions 35% of AS

Unit AS2 1 hour 30 minutes Section A, containing 10 multiple choice questions, and Section B, containing a number of structured questions. 35% of AS

Unit AS3 2 hours 30 minutes taken by candidates under controlled conditions. Section A consists of two practical tasks. Section B has a planning exercise and a number of other questions testing knowledge of practical techniques, observations and calculations. Internally assessed.

Unit A21 2 hours Section A, containing 10 multiple choice questions and Section B, which has a number of structured questions. 40% of A2 20% of A Level

Unit A22 2 hours Section A, containing 10 multiple choice questions and Section B, which has a number of structured questions. 40% of A2 20% of A Level

Unit A23 2 hours 30 minutes A practical examination consisting of a planning exercise and practical exercises. Internally assessed. 20% of A2 10% of A Level

EDEXCEL

Unit 1 1½ hours **Section A** Objective test questions. **Section B** a mixture of short and extended answers questions

Unit 2 1¼ hours **Section A** Objective test questions. **Section B** a mixture of short and extended answers questions. **Section C** Contemporary context questions

---

Grade booster

In answering Multiple Choice questions eliminate as many incorrect responses as you can.:
e.g. Which one of the following has the most covalent character.

A AlF_3
B $BeCl_2$
C $MgCl_2$
D NaBr

If you know that $MgCl_2$ are ionic solids then the answer must be A or B. You will have given yourself an even chance of the correct answer.

Don't forget

There is no penalty for a wrong answer in Multiple Choice. So, if you really do not know which response is correct, then guess.

The Jargon

All the Awarding Authorities will use short answer questions or Structured Questions.
Structured questions help you by spitting up the material into part questions which require relatively short answers.
You will find examples of structured questions throughout this book.

Watch out!

Note the time allowed for each paper. Don't run out of time!

Unit 3 Chemistry Laboratory Skills Internally Assessed

Unit 4 1hour 40 minutes 1¼ hours **Section A** Objective test questions. **Section B** a mixture of short and extended answers questions. **Section C** Data questions with data booklet

Unit 5 1hour 40 minutes 1¼ hours **Section A** Objective test questions. **Section B** a mixture of short and extended answers questions. **Section C** Contemporary context questions

Unit 6 Chemistry Laboratory Skills Internally Assessed see specification for details

OCR

F321 30% of the total AS GCE marks 1 h written paper 60 marks Candidates answer **all** questions.

F322 50% of the total AS GCE marks 1.75 h written paper 100 marks Candidates answer **all** questions.

F323 20% of the total AS GCE marks Coursework 40 marks Candidates complete three tasks set by OCR. Tasks are marked by the centre using a mark scheme written by OCR. Work is moderated by OCR.

F324 15% of the total Advanced GCE marks 1 h written paper 60 marks Candidates answer **all** questions.
This unit contains some synoptic assessment and Stretch and Challenge questions.

F325 25% of the total Advanced GCE marks 1.75 h written paper 100 marks Candidates answer **all** questions. This unit contains some synoptic assessment and Stretch and Challenge questions.

F326 10% of the total A level marks Coursework 40 marks Candidates complete three tasks set by OCR. Tasks are marked by the centre using a mark scheme written by OCR. Work is moderated by OCR.

WJEC

CH1 (1hr 30min) The paper has two sections A and B. Section A consists of short answer objective questions. Section B consists of structured questions. There will be no choice. 40% of AS

CH2 (1hr 30min) The paper has two sections A and B. Section A consists of short answer objective questions. Section B consists of structured questions. There will be no choice. 40% of AS

CH3 This unit will consist of two practical exercises that may be taken from exemplars produced by WJEC, modifications of these or centre devised tasks (using the exemplars as a standard). In the latter two cases, approval must be obtained from WJEC. 20% of AS

CH4 (1hr 45min) Two sections A and B. Section A consists of structured questions. Section B consists of two questions of 20 raw marks each which are designed to produce extended responses drawing on understanding of a range of concepts. There will be no choice. 20% of A level

Links

See page 63: Weak acids and bases

Examiner's secrets

If you find it hard to do a 3-D drawing, add a label to describe the shape in words. The ammonia molecule is pyramidal.

Links

See page 67: Buffers and indicators

Watch out!

Candidates often lose marks by omitting units and by giving a numerical answer with inappropriate significant figures: e.g. we would accept 0.088 but not 0.9 for full marks.

Grade booster

You can often simplify acid–base calculations if you remember the expression $pH + pOH = pK_w = 14$ for aqueous solutions at 25 °C.

CH5 (1hr 45min) Two sections A and B. Section A consists of structured questions. Section B consists of two questions of 20 raw marks each which are designed to produce extended responses drawing on understanding of a range of concepts. There will be no choice. 20% of A level

CH6 This unit will consist of two practical exercises that may be taken from exemplars produced by WJEC, modifications of these or centre devised tasks (using the exemplars as a standard). In the latter two cases, approval must be obtained from WJEC. 10% of AS

The Jargon

An *objective* describes what you should be able to do.
Assessment objectives describe the abilities your examiners will test.

Assessment Objectives Definitions

AO1: Knowledge and understanding of Chemistry and of how Chemistry works
Students should be able to:
• recognise, recall and show understanding of scientific knowledge; and
• select, organise and communicate relevant information in a variety of forms.

AO2: Application of knowledge and understanding of Chemistry and of how Chemistry works
Students should be able to:
• analyse and evaluate scientific knowledge and processes;
• apply scientific knowledge and processes to unfamiliar situations including those related to issues; and
• assess the validity, reliability and credibility of scientific information.

AO3: How Chemistry works
Students should be able to:
• demonstrate and describe ethical, safe and skilful practical techniques and processes,
selecting appropriate qualitative and quantitative methods;
• make, record and communicate reliable and valid observations and measurements with appropriate precision and accuracy; and
• analyse, interpret, explain and evaluate the methodology, results and impact of their own and others' experimental and investigative activities in various ways.

Watch out!

Note that the Practical units are heavily biased towards AO3.

Assessment objective percentage weightings

Knowledge and understanding of chemistry and of how chemistry works

Application of knowledge and understanding of chemistry and of how chemistry works

How Chemistry works

AQA unit	AO1	AO2	AO3	overall	EDEXCEL unit	AO1	AO2	AO3	overall
1	7	8	2	17%	1	10	8	2	20%
2	9	11	3	23%	2	8	10	2	20%
3	1	1	8	10%	3	1.2	1.2	7.6	10%
4	6	10	4	20%	4	5.3	9.4	5.3	20%
5	6	10	4	20%	5	6	10.9	3.1	20%
6	1	1	8	10%	6	1.2	1.2	7.6	10%
	30	40	30	100%		31.7	40.7	27.6	100%

CCEA unit	AO1	AO2	AO3	overall	WJEC unit	AO1	AO2	AO3	overall
AS1	42.5	42.5	15	100%	CH1	8.75	8.75	2.5	20%
AS2	42.5	42.5	15	100%	CH2	8.75	8.75	2.5	20%
AS3	20	20	60	100%	CH3	1	1	8	10%
A21	35	50	15	100%	CH4	6.75	11.25	2.5	20%
A22	35	50	15	100%	CH5	6.75	11.25	2.5	20%
A23	20	20	60	100%	CH6	1	1	8	10%
						32	42	26	100%

OCR as % of AS unit	AO1	AO2	AO3	overall	OCR as % of A2 unit	AO1	AO2	AO3	overall
F321	14	14	2	30%	F321	7	7	1	15%
F322	21	24	5	50%	F322	10.5	12	2.5	25%
F323	3	2	15	20%	F323	1.5	1	7.5	10%
					F324	5	9	1	15%
					F325	9	13.5	2.5	25%
					F326	1	1.5	7.5	10%

Data handling 1: calculations and problem solving

At A-level you will be given data to interpret and convert from one form into another. The data will usually be in the form of tables, diagrams and graphs but sometimes it may just be information described in a few sentences.

Watch out!

In this book you will find relative atomic masses, *A*r, and atomic numbers, *Z*, in the periodic table on pages 90 and 91.

Physical constants

→ Physical constants have a name, a symbol, a value and units:

Avogadro constant	L	6.02×10^{23}	mol^{-1}	per mole
elementary charge	e	1.602×10^{-19}	C	coulomb
gas constant	R	8.314	$J\ K^{-1}\ mol^{-1}$	joule per Kelvin per mole
Faraday constant	F	96 500	$C\ mol^{-1}$	coulomb per mole
Molar gas volume	V_m	22 400	cm^3	cubic centimetre

The jargon

e = charge of the electron V_m is at 273 K and 101 325 Pa

Exam question 1
answers: page 213

A direct current of 0.386 A passed through dilute aqueous sulphuric acid for 10.0 s produced a tiny bubble of hydrogen at an inert cathode. Calculate (a) the amount of charge passed in coulombs, (b) the volume of the hydrogen bubble at 273 K and 1 atm, and (c) the number of H_2 molecules in the bubble. [1 A s = 1 C]

Grade booster

Wherever possible the arithmetic is kept simple and to two or three significant figures. So you are expected to make clear your working and give answers to appropriate significant figures and with correct units.

Infrared Spectra

→ An infrared correlation table lists absorption frequency range, class of compound and group(s) causing the absorption.

When bond stretching or bending cause a molecule to absorb energy, less infrared radiation passes through the sample and a trough appears in the spectrum.

→ Infrared spectra display percentage transmission against frequency.

The jargon

Wavenumber = 1/wavelength.
4 000–2 500 single bond stretching
2 500–2 000 triple bond stretching
2 000–1 500 double bond stretching
1 500 1 000 finger print region = part of spectrum with (usually) a complex pattern that can be analysed and stored by a computer for identifying individual compounds.

Wavenumber/cm^{-1}	Class of compound	Bond
3 750–3 200	alcohols and phenols	O—H and N—H
3 095–3 010	alkene and arene	C—H
2 970–2 850	alkane	C—H
2 260–2 100	alkynes and nitriles	$C \equiv C$ and $C \equiv N$
1 740–1 720	aldehyde	C=O
1 700–1 680	ketone	C=O
1 680–1 620	alkenes and arenes	C=C
1 480–1 360	alkenes and arenes	C—H
1 200–1 050	alcohols	C—O
1 200–800	alkanes	C—C
800–600	chloroalkane	C—Cl
600–500	bromoalkane	C—Br

Examiner's secrets

You are always given infrared data when necessary and in a simplified form like the table on this page. You will usually be expected to use the information to interpret spectra and to identify compounds and their structures.

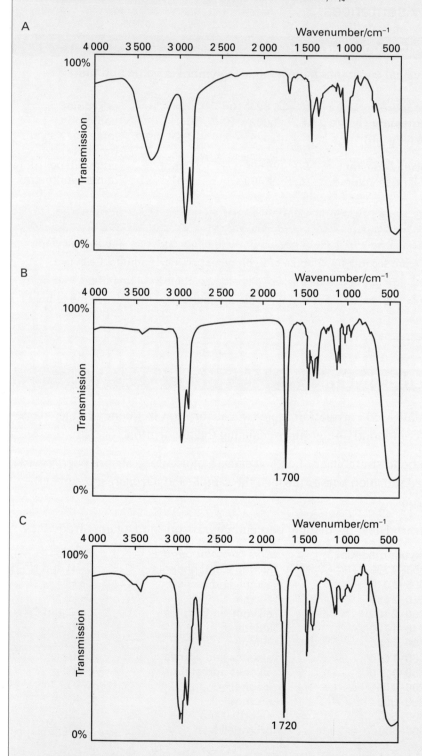

Three straight-chain organic compounds, A, B and C, have the infrared spectra shown below. One compound has the molecular formula $C_8H_{18}O$ and the other two are isomeric with the molecular formula $C_7H_{14}O$.

A

B

C

Use the IR correlation data to identify the class of compound and the functional group in A, B and C and write a possible structural formula for each compound.

Data handling 2

Nuclear magnetic resonance

→ A chart of ^1H resonances shows groups containing the hydrogen
   atom and the corresponding range of chemical shifts, δ, in ppm.

When resonance occurs (because the radiowave energy from the
transmitting oscillator matches the energy difference of the nuclear spin
states) the receiver detects a decrease in signal intensity.

→ NMR spectra display absorption against chemical shift (δ).

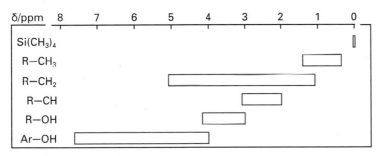

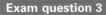

Exam question 3   answers: page 213

Two straight-chain organic compounds, X and Y, gave the NMR spectra shown
below. The empirical formula of one is C_2H_6O and that of the other is $C_4H_{10}O$.

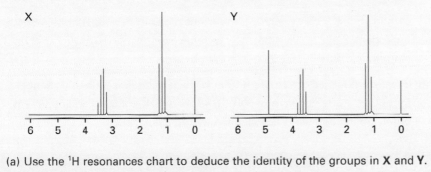

(a) Use the ^1H resonances chart to deduce the identity of the groups in **X** and **Y**.
   Write the name and structural formula of each compound.

(b) Describe one chemical test to distinguish **Y** from **X**.

Average Bond Energies

→ Bond energy (or enthalpy) is the average energy required to break
   one mole of the specified bond and is always a positive value.

→ Some values and the bond dissociation energies (for diatomic
   elements) are values applying to specific molecules.

Examiners often give the necessary data with a question but sometimes
you have to select what you need from a larger table in your data book.

Bond	Energy/kJ mol^{-1}	Bond	Energy/kJ mol^{-1}
C=O	805	N≡N	945
H−O	464	N=O	470
C−H	413	N−H	388
C−N	286	N−N	158

In N_2O_4 the bonding is delocalized and the NO bond is not strictly speaking N=O. See page 105

Exam question 4

answers: page 213

1,2-dimethylhydrazine, a potential rocket fuel, ignites spontaneously on mixing with dinitrogen tetraoxide, a potential oxidant. The following is a possible equation for the combustion reaction:

$$CH_3-NH-NH-CH_3 + 2N_2O_4 \rightarrow 2CO_2 + 4H_2O + 3N_2$$

Use the table of average bond energies to calculate the energy change for the reaction. Comment briefly on the suitability of a mixture of these two nitrogen compounds as a rocket propellant.

Standard enthalpy changes

Data books list $\Delta H^\ominus_{f,298}$ for inorganic and organic compounds and $\Delta H^\ominus_{c,298}$ for organic (and some inorganic) compounds. You may be given the necessary values (usually in kJ mol^{-1}) with the question but you may be asked to choose values from your data book.

Exam question 5

answers: page 213

(a) Use the standard enthalpy changes of combustion (in kJ mol^{-1}) in the table to calculate the standard enthalpy of formation of each of the four cyclic hydrocarbons.

carbon	C	–394	cyclohexadiene	C_6H_8	–3543
hydrogen	H_2	–286	cyclohexene	C_6H_{10}	–3752
benzene	C_6H_6	–3267	cyclohexane	C_6H_{12}	–3920

(b) Use the enthalpies of formation from (a) to calculate the enthalpy change for (i) $C_6H_6 + 3H_2 \rightarrow C_6H_{12}$ and (ii) $C_6H_8 + C_6H_{10} + 3H_2 \rightarrow 2C_6H_{12}$ and comment briefly on the difference between the two values.

Standard Redox Potentials

→ Data books list the standard e.m.f. of electrochemical cells in which the standard hydrogen electrode forms the left-hand side.

→ Values of $E^\ominus$/V are listed with the right-hand side of the cell and are usually arranged in order from most negative to most positive.

Exam question 6

answers: page 213

Electrode system	$E^\ominus$/V
X $[Cr_2O_7^{2-}(aq) + 14H^+(aq)],[2Cr^{3+}(aq) + 7H_2O(l)] \mid Pt$	+1.33
Y $Cl_2(aq), 2Cl^-(aq) \mid Pt$	+1.36
Z $[MnO_4^-(aq) + 5H^+(aq)],[Mn^{2+}(aq) + 4H_2O(l)] \mid Pt$	+1.51

Use the above data to answer the following questions.

(a) Write the complete cell diagram for the electrochemical cell with a standard e.m.f. of –1.36 V.

(b) What would be the standard e.m.f. of a cell composed of **X** and **Z**?

(c) Why do manganate(VII) titrations use sulphuric acid and not hydrochloric acid to acidify the solutions?

(d) What would be the effect, if any, of adding aqueous potassium dichromate(VI) to aqueous manganese(II) sulphate?

Answers
Data handling

1 (a) The charge passed is the time in seconds multiplied by the current in ampères.
   Charge = $0.386 \times 10.0 = 3.86$ C.

 (b) No. of moles of electrons passed = $3.86/96\,500 = 4.00 \times 10^{-5}$.
   $2H^+ + 2e^- \rightarrow H_2$
   Therefore, 2 moles of electrons form 1 mole of hydrogen molecules.
   No. of moles of hydrogen molecules formed = 2.00×10^{-5} mol.
   Thus volume of bubble (at s.t.p.) = $22\,400 \times 2.00 \times 10^{-5}$ cm^3 = 0.448 cm^3.

 (c) No. of molecules of hydrogen in the bubble = $2.00 \times 10^{-5} \times 6.02 \times 10^{23} = 1.20 \times 10^{19}$.

2 Compound A appears to be an alkanol. The absorptions correspond to alkane and alcohol. It is not possible to distinguish between a primary and secondary alcohol. A could be $CH_3(CH_2)_6CH_2OH$.

   Compounds B and C have similar absorptions around 1700 cm^{-1}. The alkane absorptions suggest that they are aliphatic carbonyl compounds. The carbonyl group absorption for compound B has a slightly larger wavenumber than that in compound C. Therefore compound B is an aldehyde such as $C_6H_{13}CHO$ and compound C is a ketone such as $(C_3H_7)_2CO$.

3 (a) Examination of the two NMR spectra shows that compound Y has a peak corresponding to a single ^1H in an — OH group. The remainder of the spectrum of Y is consistent with a quadruplet of lines for — CH_2 protons coupling with —CH_3 and a triplet for the — CH_3 protons. The spectrum is consistent with ethanol, CH_3CH_2OH.

   Compound X is consistent with the presence of a –C_2H_5 group, with the characteristic quadruplet and triplet of lines. Compound X is likely to be ethoxyethane, $(C_2H_5)_2O$.

 (b) There are a number of chemical tests to distinguish the two compounds. Ethoxyethane is classified as an ether and is relatively inert. The —OH group of the alcohol is quite reactive, e.g. ethanol reacts with metallic sodium to give hydrogen, ethoxyethane does not. Ethanol readily reduces acidified potassium dichromate(VI) to give a colour change from orange to green. Ethoxyethane does not. Only ethanol gives the iodoform reaction.

4 Bonds broken (endothermic), ΔH is positive:
   $6(C-H) + 2(N-H) + 2(C-N) + (N-N) + 4(N=O) + (N-N)$
   $= 6\,022$ kJ
   Bonds formed (exothermic), ΔH is negative:
   $4(C=O) + 8(H-O) + 3(N\equiv N) = -9\,382$ kJ
   The enthalpy change for the reaction is $-3\,360$ kJ mol.

   This reaction is extremely exothermic and produces a large volume of gaseous products. This is ideal for jet propulsion since a large volume of hot gases is produced from a relatively small volume of liquid reactants.

Examiner's secrets

From the information given in the table, you must assume a structure for N_2O_4 which is not strictly correct. See page 105.
   Niceties of chemistry are sometimes ignored to set you a more straightforward problem.

5 (a) Applying Hess's law to each of the following,
   $C_6H_6 + 7\tfrac{1}{2}O_2 \rightarrow 6CO_2 + 3H_2O;\quad \Delta H_f = +45$ kJ mol^{-1}
   $C_6H_8 + 8O_2 \rightarrow 6CO_2 + 4H_2O;\quad \Delta H_f = +35$ kJ mol^{-1}
   $C_6H_{10} + 8\tfrac{1}{2}O_2 \rightarrow 6CO_2 + 5H_2O;\quad \Delta H_f = -42$ kJ mol^{-1}
   $C_6H_{12} + 9O_2 \rightarrow 6CO_2 + 6H_2O;\quad \Delta H_f = -160$ kJ mol^{-1}

 (b) (i) $C_6H_6 + 3H_2 \rightarrow C_6H_{12};\quad \Delta H = -205$ kJ mol^{-1}
     (ii) $C_6H_8 + C_6H_{10} + 3H_2 \rightarrow 2C_6H_{12};\quad \Delta H = -313$ kJ mol^{-1}
   In both (i) and (ii) three double bonds have been hydrogenated. Less energy is given out from benzene because the molecule is more stable due to delocalization.

6 (a) $Pt\,|\,Cl^-(aq),\ Cl_2(g) \vdots\vdots 2H^+(aq)\,|\,[H_2(g)]Pt$

 (b) $1.51 - 1.33 = +0.18$ volts

 (c) Since the standard redox potential of the manganate (VII) system indicates that the oxidation of chloride ions to chlorine is feasible, sulphuric acid is used to produce aqueous hydrogen ions so that no manganate(VI) ions are involved in oxidation of chloride ions.

 (d) There would be no effect, since the total e.m.f. for such a cell would be negative and the reaction would not be feasible.

Index